INTO THE WOODS

THE BATTLE FOR TASMANIA'S FORESTS

Anna Krien

Black Inc.

Published by Black Inc.,
an imprint of Schwartz Books Pty Ltd
22–24 Northumberland Street
Collingwood VIC 3066, Australia
enquiries@blackincbooks.com
www.blackincbooks.com

9781863955584 (paperback)
9781921870545 (ebook)

NATIONAL LIBRARY OF AUSTRALIA

A catalogue record for this
book is available from the
National Library of Australia

Book design by Thomas Devereall
Typeset by Duncan Blachford

Contents

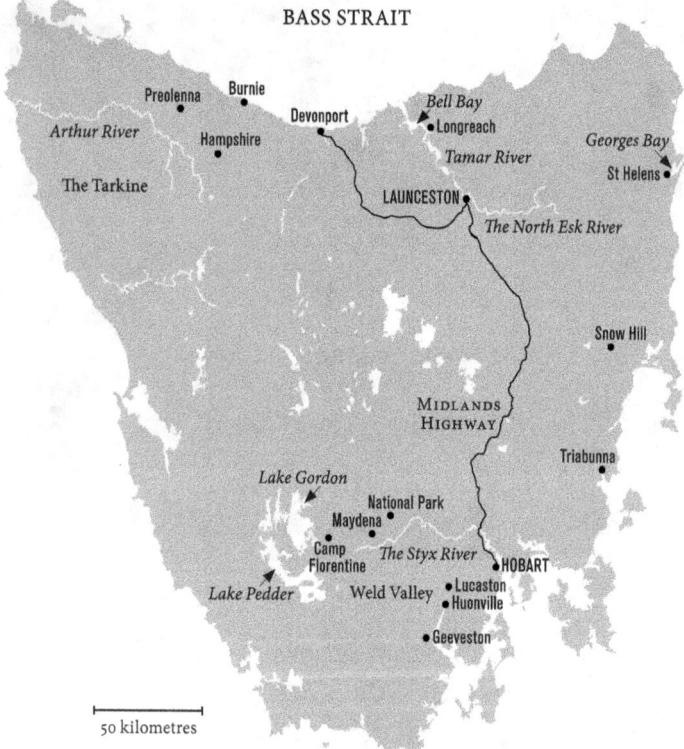

BASS STRAIT

Preolenna •   Burnie •

Arthur River   Devonport •   Bell Bay •
                                    • Longreach

              Hampshire •   *Tamar River*   *Georges Bay*

The Tarkine   St Helens •

                          LAUNCESTON •

                              The North Esk River

                                        Snow Hill •

                    MIDLANDS
                    HIGHWAY

                                        Triabunna •

        Lake Gordon

                    National Park •
              Maydena •
         Camp   *The Styx River*
         Florentine •   • HOBART

Lake Pedder   Weld Valley   • Lucaston
                              • Huonville

                    • Geeveston

⊢———————⊣
50 kilometres

iv

PROLOGUE

Sun in my eyes, I'm taking the corners too fast. Slow-moving swarms of insects swallow the road whole and I realise I'm going to cry. Log trucks coming the other way pin me in their breach. Like a magnet, my car resists them, hugs the cliff, then surrenders – sucked into the hauler's back draft. This morning I placed my hand on the black fur of a dead devil. Still warm, the little fellow was not yet frozen by the frost. A white V of fur across his chest looked like a football jersey. In the corner of his mouth a tear of blood was the only sign that the dead was in him, not yet reverberating into his pelt, his floppy legs.

The man who showed me the dead devil had let me sleep the night in an old asbestos shack on his property surrounded by paprika-coated boulders and giant leather flanks of kelp and ocean. On our way there he picked up a dead wallaby from the side of the road and, when we reached the turn-off to the shack, cut its belly open with a Swiss army knife and lassoed its legs with rope to the bull bar. Dragging the roo over the button-grass plains, its guts oozing out, Geoff hoped the scent of blood would lure the devils off the sealed road, away from cars and the thugs who sometimes aimed for them.

Geoff's one good eye glints. The other is permanently glazed with a film of blue-green glass. After half an hour of undulating velvet mossy humps, we arrive at the shack balanced on the north-west corner of Tasmania, in the direct path of the Roaring Forties, westerly winds that whip around the earth. Inside, empty wine bottles with candles in them crowd the mantlepiece, a faded surfing poster and whale migration maps are blu-tacked on the walls, and a mirror flecked with rust from the sea salt is propped against the wall. I roll out my bedding next to the pot-belly stove as Geoff stakes the dead wallaby outside the window so we can watch to see if a devil has followed the scent.

In the last light we go for a walk along the rocks.

'Look down,' he says. Around my feet is a scattered collection of stones and shells – periwinkles, abalone, mussel, welk and those tiny white shells that look like ears. I shrug at him in ignorance. 'It's a midden,' he tells me. 'The remnants of an Aboriginal feast.' I look closely and notice the stones are sharp on one side and smooth on the other. He points to a circular depression in the ground and then another: miniature craters dot the coast. 'Huts.' The giant thumb-prints paved with white rocks were the homes of Aboriginal families, who built wooden ribcages to bow over the kneaded earth, lining the huts with kangaroo skins. 'There's another just near a seal hide, a place where a seal would belly up for a snooze and be killed for tucker.'

'We've had this land for generations,' he muses. 'But it's like I never saw it.' He's right. The middens I walked over without thinking now catch my eye, the sharpened stones like abandoned cutlery from another time. Geoff used to run cattle on this land, as his family has since the late 1800s, but he removed them once he started to understand the effects of hoofs on the soil. But he still hasn't been able to stop the trail bikes and four-wheel-drives from carving up the landscape. A splay of exposed roots reveals a series of broken dunes and pulverised middens. Down the beach cows are standing on the sand, their leathery skins blasted with sea-spray.

'About four years ago, about a thousand locals met at the river to protest against turning the area into a national park,' Geoff says. Most of them arrived on trail bikes and in four-wheel-drives, converging on the mouth of the Arthur River in the Arthur-Pieman Conservation Area (the Pieman being another river in the district, named for a convict who was jailed for selling pies filled with rotten meat). Geoff's brother was one of the main organisers of the protest. 'That's how it is in Tasmania,' he says, shrugging.

The ocean grumbles and it's night suddenly. At the shack, Geoff shows me how to turn the lights on and off by squeezing and releasing the clips on a battery. The thump of paws suggests the dead wallaby already has a visitor. We go to the window. A devil is here. She is sleek and stocky, wet pink nose sniffing the air and front canine missing. Geoff has hidden a baby monitor out in the grass so we can hear her; its electronic red eye blinks in the dark. A fluorescent spotlights the dead wallaby, staked out like a Jesus. I suspect the devil knows it is a set-up, but meat is meat. With one last look around she digs in, a blotch of blood on her nose.

When Geoff leaves, slowly steering the ute over the bulging plains, I settle into my swag with the baby monitor crackling beside me, its plastic partner outside. I listen to the soft pad of leather paws, the wheezy whistle through the nostrils of a feeding animal, and the bone straining, like stepping on a branch, until it snaps in the devil's jaw. Pine pops in the stove belly. Hand clasped between my thighs, I try to stifle the urge to pee. It's no good. Sighing, untangling my arrangement of blankets and sleeping bag, I step into a night splintered with stars. I can feel the scuttle of animals, their baited breath as I squat, trying to direct the hot stinky piss away from my feet, my bum standing out like a satellite dish. Back inside the shack, I feel the silence from my interruption give way as the devil's hunger returns. The sound of her, and the overhead calls of pied oystercatchers, follow me to sleep.

In the morning as I boil water on the stovetop for a cup of black tea, Geoff tells me he found a devil on the road, dead: 'Number eighteen for the year.' I imagine it printed on the devil's black and white back. He carried it to his car, laid it down in the tray and continued on his way to pick me up.

While he puts right whatever I put wrong in the shack, I splash the rest of my tea into the sink, roll up my bedding and head out to the ute. Looking at the dead devil, I pat him like a cat. He's perfect, Geoff says, coming up behind me. 'A perfect male. About two years old, ready to breed. Not a single scar on him.'

After Geoff drops me at my car, I drive all day. I don't think about the devils or the salt-licked shack. I listen to news radio, about the protests in Iran and tax-dodging ministers in Britain. There is a great big crack in my windscreen and it's travelling sideways. I wonder how long it will take before I get pulled over for it. Then, as the day wears on, the log trucks stressing me on the narrow corners, I see a black and white bundle beside the road, lumpy and broken like someone had tried to turn it inside out. I turn the car around and go back to take a closer look. Nudging it with my boot, I roll it over. Jaws yawn open in a permanent hiss.

It is a black and white moggie. And finally I cry.

RATBAGS

THE CROSSING

I am on the *Spirit of Tasmania*, a floating RSL and car park, heading south. A 290-kilometre power cable lies eel-like beneath us on the ocean floor, an extension cord drawing power from Tasmania's dams, feeding it into the power grid of my home city, Melbourne. A tourist map displayed on the wall is colour-coded in varying shades of green, specifying different types of wilderness: the remote, the near and the friendly – enough nature to satisfy any type of visitor.

There are no asterisks next to these categories in the legend, no fine-print disclaimers explaining the ongoing environmental battles behind these carefully drawn lines. It is a map, after all, not a government notice or a campaign poster. 'How would you like it if almost half of your state was locked away?' a timber-industry spokesman had said bitterly to me over the phone a few days earlier. I had asked him about logging ancient forests – if the practice would end on the island.

'The Natural State,' I read on returning vehicles' number plates as I boarded the ferry, the Tasmanian state slogan accompanied by a picture of the now-extinct Tassie tiger peering out from beneath the exhaust pipes. On the rear windows of almost half the cars I see, there are stickers in eternal argument with one another. 'Tasmania: The Corrupt State' and 'Save the Styx' versus 'Greens Tell Lies,'

'Greens Cost Jobs' or simply 'Green Scum' – slightly tamer versions of older stickers that read 'Keep Warm This Winter: Burn a Greenie.' It is said each glut of car stickers in Tasmania signals a new chapter in this intense and deeply personal debate that has been going for forty years. In the '80s, some police doffed their caps to protesters as they arrested them, revealing 'No Dams' stickers underneath, showing their allegiance to the Franklin River blockaders. Today the stickers read 'No Pulp Mill,' directed at the timber company Gunns, the largest native hardwood-chip exporter in the world, or 'Protect My Habitat,' accompanied by an image of a Tasmanian devil, now facing extinction from a disease whose cause no one can agree on.

The ship lets out a low honking moan along the esplanade and starts to roll forward over the bay. I gather my stuff and walk around the deck. At the reception a man in a wheelchair is accusing a staff member of discrimination. There's a bar at each end of the floor, the smell of bourbon and coke, beer and salt-and-vinegar chips already hovering. The 'brrrrr,' 'brrrr-ing' and 'ca-ching' of lit-up pokie machines sing out from behind frosted glass.

I put on my blue tartan lumber jacket and head outside into the misty rain and lean over the railings. Seagulls are swooping into the ferry's backwater, flushing out tumbled and dizzy fish. I watch the foggy outline of St Kilda pier, the thin grey curve of the West-gate bridge and candy-caned smelters of Williamstown disappear as we head towards the lips of Port Phillip Bay, parted by a breath of water. Three lighthouses will guide the ferry between the reef, over the rip and out into the ocean. Fewer than two hundred years ago this only entrance and exit to Melbourne was a death trap, but tonight on a ship that looks like a lit-up lounge room with people watching telly, eating dinner and leaning into one another on couches, we barely register the rocky gateway. I stare hard at the water, trying to will something to the surface: a fin, a flipper, a flying fish even. It's hard to imagine there's a world under there, beneath the choppy blue surface.

I wonder if there was ever a time the horizon didn't look so empty. I've read that the British settlers had to carefully row their boats across Tasmania's Derwent River to avoid the accidental whack of a whale fluke snapping their boat in half, or the odd roll flipping their hull. On beaches it was normal to lean against the enormous ribcages of these creatures, bleached white and porous with time. Down at the very bottom of Tasmania, facing Antarctica, lighthouses are posted on fragmented chips of land to keep an eye out for icebergs, turquoise chunks of ice detected by satellites. A little over a decade ago ships were put on high alert after an iceberg 300 kilometres long and 40 kilometres wide floated past. But so far, a good half hour into my twelve-hour journey, nothing.

I was seven years old when I first visited Tasmania. I remember four things: frozen puddles, a colourful crocheted rug, screaming like a miniature Rain Man when chickens surrounded our hire car, and vomiting. I'd say my brother and I left half of our body mass on that island, each mountain turn turning us a lighter shade of yellow. Sometimes only one of us needed to be sick but just the sight of Mum holding one sibling over a patch of grass on the side of the road would trigger the other. Our eldest brother, relegated to sitting in the middle, fumed as we gasped at the open windows, clutching vomit-stained towels. But I also remember it as the stuff of children's books: old farmhouses, ghost stories and fireplaces, attics and creaky beds piled high with blankets, mornings when you could shimmy through the icy grass, the crisp blue horizon outlined by a thread of sun and my brothers and I pretending to smoke cigarettes, breathing out puffs of cold air.

Today a very different, more dramatic impression of the island is returned to the mainland by writers and filmmakers. It is a gothic place with a bloody undercurrent, where behind every magic faraway tree is a logger kicking in the head of an activist, where insular and genetically murky communities have been forcibly separated by authorities. A place where an angelic-looking man with blond hair and blue eyes, who took round-the-world flights just to talk to the captive passengers sitting beside him, one day picked up a gun and killed thirty-five people in an afternoon. Journalists bring back stories of threatening or confessional phone calls in the middle of the night, of great wads of cash found in a local media mogul's freezer, another wad in a premier's bedroom, a minister who thought it would be hilarious to hold a starter's pistol to a female reporter's head. Of a place where Exclusive Brethren and pig farmers alike fund thousand-dollar advertising campaigns, and where a 2000-year-old protected tree is axed, drilled and filled with diesel before being spray-painted with the words 'Fuck You Greenie Cunts' and set alight.

'It's nice out here, isn't it?' a voice says behind me, parts of it disappearing in a gust of wind. A skinny guy dressed in striped thermals is standing on the deck with a deaf woman he introduces as his German girlfriend. Introducing himself as 'Crazy John,' he tells me that they are going rock climbing in Tasmania. He demonstrates his upper-body strength by hanging off the doorframe with one arm, his other hand punching the air in excitement.

'That's amazing,' I say obligingly.

He then demonstrates the double click of a badly mended finger.

'I'll still climb, though. Even when it was swollen, I climbed on it.' Again, he shows me how he has to grip rocks with the top part of his fingers: 'That's the hardest bit. Remembering to use your finger strength not your hands.' I nod, feeling like I'm in a polar-fleece version of *Point Break* or *The Fast and the Furious*.

'Hey, so who do you know in Tassie?' he asks. 'I used to live there, you know.' Stupidly I tell Crazy John I'm planning on writing about Tasmania's forests and am heading out to the Florentine blockade. 'Ha! I was part of that! I helped set that up. I was part of the Weld actions too. It's a small world, ain't it?'

I shrug, thinking, well, Tasmania's a pretty small place.

'You should ask me some questions.'

'Such as?'

'Well, you're the journalist.'

But I'm not sure if I want to talk to Crazy John. I don't know how it happened, but he's cornered me on the deck. He bums a cigarette off a nearby ferry worker having a smoko and sucks it in. 'I just climb now. You'll see. Those hippies don't do nothing – it's all about peaceful protest and shit. Besides, I've done my bit. I'm sterilised. It's up to you breeders to do something about it now. You breeders gotta put some work in … Future primitivism, have you read about that? … They don't do anything, at a coal protest they just stood there. Then me and this other guy just picked up the fence and threw it on the ground …' And so on. Even his deaf girlfriend gets bored and goes inside. I decide I hate Crazy John. 'It's all about veggie patches. I mean you wouldn't understand how great it is to have your own veggie patch, people need to know where things come from.' He sucks the last centimetre of tobacco to a stick of ash and flicks the butt into the sea. Watching the butt arch and disappear into the ocean, I can't believe it.

'You've got to be kidding me,' I say.

'What?'

'You just flicked your butt into the sea.' Crazy John stops and stares at me. Slowly he says, 'Turn around, man, look up there.' He points to the spout spewing out smoke on top of the ship. 'Keep your mind on the big picture,' he says. Part of me wants to cry. I had been quite happy tucked into my quilted jacket and watching the water until this fucker in thermals came along. I start to walk away.

'I actually just did it to test you,' he calls out after me.

'Well, you better jump in and grab it now the test is over.'

'No, I wanted to test if you were a big-picture person, or a writer getting stuck on the little things.'

'I'll see ya on the island,' I say, walking away.

Two and a half sleeping pills later, I'm out cold. In the morning, I wake up to the sound of a familiar registration number being called out over the loudspeaker. Groggily I run downstairs. My car is alone in the floating car park, a couple of workers standing listlessly around it. I bunny-hop off the boat and onto Tasmania.

MIRANDA AND NISH

'GET OUT OF MY WAY YOU FUCKEN FEEERALLLLLS, YOOOUUU FUCKEN BAHHHSTARDS, GET THE FUCK OUT YOU CUNTS I WARNED YA I FUCKEN WARNED YA GET YOUR FUCKEN ARMS OUT GEEET YOOURR FUCKEN ARRRMS OUT YOU FUCKEN CAAAUUNTS.'

Miranda and Nish huddle under a blanket; they can hear the windows being smashed and glass falling on top of them. Hands are grabbing at their legs and pulling. Above them the car is buckling under the weight of something heavy, an axe or a crowbar maybe. More glass scatters, this time spilling like marbles under the blanket. In small flashes of light they can see the hands reaching in through the broken glass, fluoro orange arms and fat fingers snapping. Their arms are down two pipes in a cement block beneath the car, locked fast to metal bars inside. Miranda feels Nish lock off. She does the same.

'I WARNED YA I FUCKEN WARNED YA I WARNED YA I FUCKEN WAARNED YA YAAA FUCKEN CAAAUNNNTS ...'

Nish is trying to get out and they pull him the other way, towards them, dragging him through the broken glass; they've got a hold of his leg, the blanket is going with him. Miranda grabs it and pulls it back over her. His shirt is up round his neck. He falls to the ground, gets on his hands and knees. He's kicked in the head. Not as hard

13

as he thought he would be. Kicked again. Nish scrambles into the bush, tripping up into the trees.

'GET YOURR FUCKEN ARM OUT GET FUCCCCKEN OUT ...'

They keep smashing the car, yelling at the second body inside. Miranda flings the blanket off, seeing only the jagged edges of the rear window, fluoro orange blur, she falls over the boot and onto the road. Her hand is bleeding. She can't run. Legs jelly. Walks slowly away.

'RIGHT. NOW LET'S GET TO FUCKEN WORK. SOMEONE PULL THIS FUCKEN THING OUT OF HERE.'

Tension sucked out as fast as it came in, the men walk away, orange shirts nudging one another, younger ones smirking at the shell-shocked activists standing lamely to the side, like sightseers at a very disappointing parade. The men make stabbing motions at their necks.

'We'll get you.'

'If we see you again, we'll fucking kill you.'

'You're dead.'

The loggers, however, are unaware of a girl, about eighteen years old, who clambered barefoot up a tree when they arrived, video camera slung over her back, and lay on a branch, like a cat out of *The Jungle Book*, to film them.

*

It was this footage that made me open another window on my computer and book a seat on the ferry. I know one of the girls living at the blockade, Ula Majewski. Many months before the attack I saw her at a festival in Newcastle. Usually dressed colourfully, like a parrot, this time she had been subdued, her flamboyant feathers muted. She wore a grey coat speckled with lint. Her blonde hair was matted into dark sandy dreadlocks and she had lost weight. The city blur of eating out, drinking and partying had disappeared from her cheeks and neck, sharpening her jaw-line. Her watery pale blue eyes had hardened into chips of turquoise. In the evening I danced over to her and tried to drag her onto the dance floor. But she shook her head, untangling her hands from mine. Later, she said to me, 'I shouldn't be here. I need to get back to the Florentine. I shouldn't be here.' Then she looked at me with such intensity I had to look away. At the end of the night, walking with friends back to our hostel, one of them asked, 'What's up with Ula?' I said I didn't know and honestly, I didn't.

Ula has been part of the Upper Florentine blockade for three years. This live-in protest in the south-west of Tasmania was first set up in 2006 when the old-growth forest's number came up for logging in the state forest agency's production schedule. A few locals constructed tree-sits and tents to block the road into the valley, but their numbers needed bolstering. According to some of these original temporary inhabitants, the 'ferals' were reluctantly permitted to move in. Aside from an early bust at the beginning of 2007, when a section of the road was gravelled and sixteen activists arrested, the blockade has been mostly left alone. Of the ten coupes, two have been logged. 'Working' forests are divided into coupes and then identified with codes such as SX10F. Ula says these labels have the effect of distancing the logging operations from the reality of levelling forests. Most coupes are between 40 and 100 hectares.

When I ask Ula in an email what it is like to live at the blockade, she writes back: 'Directions to a friend's house become … "Chuck a right at the big myrtle, go down the hill a bit, jump over the fallen-down stringybark with the mad orange fungus on it, and you'll be able to see the tarp just behind the ferns."' She described the daily rituals of camp life: carting fresh water from the river to the makeshift kitchen, digging 'shit pits,' inhaling the smell of bruised leaves after a storm. 'I have to rediscover my "forest legs,"' she writes on her return from the mainland, her sloppy heavy-footed city walk tripping her up and over logs.

Three days after the sledgehammer attack on the car and two evenings before I arrive, the Florentine blockade is visited around midnight. The blockade was already on a 'red alert' after the attack on the car. Scouting missions were reined in. The all-seeing bird's nest positions in the canopy were occupied in shifts and activist change-overs were extra careful, no one locking off until certain their replacement had locked on. So when the three carloads of men arrived, parked so their headlights and kangaroo-hunting spotlights shone into the mouth of the forest, the siren had sounded ten minutes earlier and everyone had bolted.

Almost everyone.

'Wakey, wakey,' one man called out in a singsong voice as they stepped into the blockade, holding tools and lugging jerry cans full of petrol. A deep sleeper was still curled up in his sleeping bag in the back of a car. An axe blunting into the roof above him did the trick. Scrambling to the other side of the car, he fell out the door and half-ran, half-crawled into the bush. They didn't follow him into the forest. Some activists climbing trees, others crouching beneath ferns, they stayed quiet as the men taunted them to come into the open. From tree-sits high up in the canopy, three protesters peered over their wooden platforms, quietly checking their cables and testing the ropes. The visitors started to swing their hammers and axes into the activists' cars, smashing the windows and clubbing the doors. They pushed over a small information hut, knocking the 'Save the Florentine' pamphlets into the mud, slashed

16

the banner they could reach and then doused the area with petrol. Flicking a handful of lit matches into the dark, they got back into their cars and drove away. And as the flames licked up the petrol, the hiding protesters slowly emerged, calling out to one another and up to the people in the tree-sits, checking to see that everyone was okay. In a huddle they watched the burning vehicles and hut, jumping each time something – the fuel tanks, gas cooker, bottles of cheap cider – exploded in the heat.

THE PINK PALACE

I've been admiring the front gardens with their roses, artichokes and cascading yellow flowers on the steep walk up to the Pink Palace, a shared house in South Hobart in the shadow of Mount Wellington. In the distance I can see the rubbish tip with its eternal mobile of seagulls bobbing over it. At a blue picket fence leaning drunkenly onto the footpath, I stop. The house's pink weatherboards look as if they are bursting at the seams and a giant rat's nest of pillows, blankets and lengths of foam has formed on the verandah. The odd brown dreadlock sticks out like a tail.

Stepping over the sleeping bodies, cables, axes, saws and D-locks, I walk tentatively through the wide-open front door. I feel like calling out 'Avon Calling' in a singsong voice, but shyness gets the better of me. Wandering the entire length of the house, I'm spat out into a backyard strewn with newspapers, rotting fruit, empty cider bottles, ashtrays and fermenting tubs of Yakult. Some guys are parked on a couple of corduroy couches; the cushions are ripped open, and there are telltale clumps of foam around the jaws of two wrestling staffies. Beyond the chaos is an orderly veggie patch, healthy and green under a mulberry tree.

'Too many potatoes,' one of the guys remarks when I comment on it. 'Bloody Wazza, all he wants to plant is potatoes.' Past the veggie patch is a cubby house. Someone's sleeping bag is poking out of it.

Another of the men introduces himself as Nathan, then tells me Ula Majewski is out with reporters from *Stateline*, showing them the blockade and the remains of the smashed camp.

'They said they wanted to get a human angle,' he explains. I roll my eyes.

'Of course they do,' I say, plonking myself down on a spare couch. After a pause, I admit I said the exact same thing to her on the phone.

Nathan laughs. 'Yup. That's what she told me.'

When I ask them about Crazy John, no one seems to know who he is. 'He says he knows you guys,' I say, but they shake their heads.

It's late afternoon when Ula gets back; I've been killing time in the backyard waiting for her. We hug.

'Anna Krien,' she says teasingly, emphasising each syllable. 'You're such a journo. You only come down when blood spills.'

I shrug. 'Of course. Question is, will there be more?'

I take a step back to look at her. She is looking stronger than when I saw her last, happy to be back in the thick of this chaos. Her shoulders are broad, no doubt from paddling into the surf and catching waves. Even here, where the ocean is icy all year round, she dons a steamer and slips like a seal out beyond the breaks. Five years ago, she received a scholarship to do her masters in environmental management at the University of Tasmania. She now calls the island home. In Melbourne she was a rainbow-wearing, beer and red goon-swilling student and poet. Now she is a jeans and T-shirt wearing, beer, cider and red goon-swilling activist. She is also, according to the forestry industry, an eco-militant, Trotskyist, industrial terrorist, guerrilla warrior and radical fundamentalist.

'Oh, and a slut,' she says. 'Don't forget slut.'

'Yeah, but you were a slut in Melbourne too.' We laugh. Ula has a guttural laugh. It seems to scrape up and out of her chest, her Adam's apple bobbing up and down her tanned neck. She squirts wine from a cask into two chipped mugs and hands one to me. We sit down on the back step and Ula pulls out a worn leather pouch. She rolls a reedy cigarette, cups it, lights up and inhales. The house is filled by the sound of a whooping cough; someone stops what he is doing and steps abruptly into the garden to loosen the phlegm in his chest.

'The Pink Palace is the town base. We make sure the blockade isn't forgotten,' Ula tells me. The past few days have seen videotapes and memory cards passed from one bombed-out car to another, travelling from the forest to the city so the Palace crew can churn out media releases.

'But it's not usually like this,' Ula tells me, her phone going off again; it is news radio asking for a quote. When she has finished talking about the 'industrial-scale devastation wreaked on these ancient forests,' she hangs up and tells me it took a bit of convincing to persuade her that they ought to take the footage of the violence to the media. 'But after the camp was burnt out, I knew we did the right thing. There needs to be a record in case anything worse happens.'

Many of the core group prefer to produce reports, survey coupes and release their own research to the media – 'which gets the journos excited,' Ula says sarcastically. Later she shows me some of the work the group has produced. There are flora and fauna surveys, night footage of endangered species, marsupial hair samples, coupe documentation as well as two major reports, one done in conjunction with the Wilderness Society on the island's export of old-growth timber, the other a report to the World Heritage Committee that helped culminate in a UNESCO recommendation that the Upper Florentine be protected. I'm amazed at the amount of work and at the 'matter-of-fact' style the reports are written in. It's unexpected from this motley mob, who call themselves Still Wild Still Threatened (SWST).

In the past eighteen months Ula has found herself, as she puts it, the group's 'media slut.' With her blue eyes and blonde hair, she scrubs up well for the cameras, and her easy-going and reliable manner seems to have won over the local reporters. Her key role in the group's research as well as her own study means she is a walking information booth on Tasmanian politics, local forestry practices and the environment. But the exposure in a place as small as this has a toll. She has received threats and all manner of hate mail. Her post, she says, is opened regularly. 'But it's mostly letters from my mum. She sends me clippings from newspapers on the mainland.' When I ask if there is any method to deciding who does what, she shakes her head. 'Only madness.'

Every fifteen minutes or so the back gate creaks open and people lug in boxes of dumpstered food, creating a skirmish among the dogs. There is an endless rolling of cigarettes, ruffling of newspapers, accidental knocking over of abandoned mugs and glasses. In the kitchen, a couple of women stir pots of soup and dhal, which will be taken down to the city's main park to feed the homeless and anyone else who's hungry. A seemingly mongrel mix of pirates, ferals, punks and hippies, the SWST crew are protégés of the '70s without the rainbows or earnestness. While many of yesterday's hippies have settled into invisibility on hobby farms in the hinterland or in the coastal suburban sprawl, these ratbag ferals are upfront and in your face.

'Fucking useless hippies,' one mutters, moaning about the arrival of rainbow people at the forest blockade. 'They're doing a mega-meditation sit-in on Timbs Track!'

Patting him patronisingly on the back, Ula cooes, 'Don't worry, the weather will take care of them. One cold snap and they'll be gone.' This is a running joke, I soon discover: baiting the older activists.

'This is not about some spirituality hippy shit,' one of the younger activists says to me later. 'It's deeper than that.' And if ferals are on the run from the mainstream, it's their vermin status, unwashed clothes and smelly feet that succeed in keeping the market at bay. The 'Feral Cheryl' doll is about as close to merchandising the culture has ever come. Made in South Australia, the doll came with her own stash of oregano, dark thatches of pubic and underarm hair, piercings, tattoos, homemade clothes and dreadlocks tangled with feathers and beads.

The next morning I wake up when Chops, one of the dogs, pushes me off my mattress with his boxy head and takes my blanket. Ula has set me up on the floor of her room. 'No one is allowed to crash in here,' she said as she showed me in. 'This is my sanctuary. But you're fine.' I'm glad of the special treatment, despite her mobile phone *ribbet*-ing like a frog as text messages arrive throughout the night. In the morning, I gaze sleepily at the titles of the books stacked up against the wall. Richard Brautigan's *Revenge of the Lawn* catches my eye and I start to flick through it until the 'cock-a-doodle-doo' of Ula's alarm clock urges me up. By the time I wander into the hallway, my bare foot stepping on a lint-covered bone, all the main nuts and bolts of the Palace are up. Ula has pulled her dreadlocks into a ponytail and is using a dishcloth to scrub stains off her suit pants, a cigarette burning out beside her.

Warrick Jordan, known as 'Wazza,' is on his way to uni – he will meet Ula after his tute to follow up on media – while Christo and Nathan, a.k.a. 'Lunchbox' ('coz he's packed like a lunchbox'), are getting ready to go fishing. As Ula speeds out the door, memorising sound-bites, she gives me a quick hug and says someone called Bridget will pick me up.

*

I have a sleeping bag stuffed with blankets, a packet of mixed nuts and wads of loo paper. Outside, Bridget stands beside her car and waits for me. Hands on her hips, black tiger stripes tattooed on her arm, unlaced combat boots and white skin with the fuzz of a peach, she shifts things around in the backseat for my gear and gives me a smile. The little yellow car smells like mangoes.

'I found a box of them in a dumpster last night,' she tells me.

A huge fan of mangoes, I like Bridget instantly. Twenty-one years old, she grew up in a small town down south near the Weld Valley, a part of the world where orchards and roadside fruit stalls give way to vast plains and towering forests. She describes shag-piles of moss, fluoro fungi, the slow amble of wombats across the grass. Sinkholes gape like mouths and icy water pours into a dark underworld of caves. When Forestry Tasmania, the state's official forestry body, announced plans in 2000 to build a series of bridges over the rivers to access more wood, Bridget was one of the first activists there.

Bridget is dreading the day when she might face her old school-mates in the forest. 'I'll have to look them in the eye knowing they left school at fifteen and don't know any other job.' The youngest of five siblings, Bridget also left Huonville High at year nine, ironically because she wanted to finish school. 'I did the rest of my schooling by correspondence,' she explains. 'I just couldn't go there anymore. I mean, I could cope with being different – but the school, it depressed me.' Bridget watched girls younger than her get pregnant and drop out, while boys whose voices had barely broken went to work in the bush. Many of those remaining had little desire to leave Huonville and planned on making the easy transition from school to welfare. Fiercely independent, she moved to a flat in Hobart and enrolled in a state program that is now defunct. 'I was given my own tutor and we studied one-on-one for the next three years. I loved it.'

Today many of her peers are better placed financially than she is. 'Yeah, they'll have homes and cars,' she says without a hint of envy. 'But they'll also have very few options. Which is why some of the guys get so angry, I think.' For many in the town, logging is their lifeblood. When a local boy gets married, a wedding convoy of logging trucks may circle the town's only roundabout, tooting their horns. When I eventually visit the town I meet a local woman, Lou Geraghty, who once copped an egg-and-bacon McMuffin in the face from a passing hauler after she managed to stop the clearfelling of native forest on private land around her home.

Bridget is driving me to Camp Florentine, the blockade about an hour and a half west of Hobart. As we drive, log trucks pass us from the opposite direction. Looking at their trays in her rear-view mirror, she identifies each load of wood. 'Old growth … sawlogs … woodchips.'

'You used to see a giant tree loaded up on a truck going through town,' she recalls. 'A single rider, they call it. Trunk the size of a petrol tanker. But you won't see that too often anymore.'

Bridget's arms encircle the steering wheel as she concentrates on the road. The black stripes on her arm are copied from the pelt of a thylacine, or Tasmanian tiger, the island's mysterious marsupial dog. Black hair tumbles around her pale face. She is quiet, but not in a shy way. There is a calm in her manner, a steadiness. When we stop at a service station, I flick a ten-dollar note out my window to her. She takes it and gives me the finger playfully, her long legs clopping along.

We drive past hops crops that look like lime-green feather boas tossed onto wire lattices. A flurry of chickens and roosters clucks off the bitumen into the gutter, while gangs of black cockatoos squawk overhead. As the yellow car lifts us up above sea level, skullcaps of snow appear on the mountaintops.

In a few months' time, Bridget will be placed under virtual house arrest for locking herself to a log truck as it comes out of the Upper Florentine Valley. After a day of alert policing – thirty officers running in constant lines alongside the machinery – there is a lull, a misjudged assumption that that's it for the day. Nearing knock-off time, one truck loaded up with wood is left unattended, and Bridget leaps in front of it as it pauses at the mouth of the road, locking herself to its bullbar. It takes two hours to cut her loose, the inspector raging all the while at his sergeants for taking their eyes off the ball. Subsequent bail conditions require her to remain housebound from 7 a.m. to 3 p.m. every day until her sentencing. Other protesters are forbidden to enter any land or premises under Forestry Tasmania's control, which amounts to about 22 per cent of the state.

But today, neither of us knows about her future confinement. She drives, talking happily, while I eat mangoes.

*

Most people travelling through Tasmania will never know of the long-running game of hide-and-seek taking place in the labyrinth of logging roads beyond the bitumen. Sightseers walk among 300-year-old trees, some of them 90 metres tall, in the Styx Big Tree Reserve; chainsaws can be heard in the distance. The road into this attraction is lined with stage-sets of wilderness. At the rise of a hill, just before the nose of the car tilts downwards, passengers might glimpse a balding peak, a fleeting insight into the world behind the verge.

They might also catch a glimpse of a bright orange Datsun before it veers off the main road and scuttles away over potholed dirt roads, followed by its equally ratty mate, a red jeep or a little yellow hatchback perhaps. They bounce down the road like packages that have been thrown, their contents threatening to burst out at any second. On the inside, their passengers are ankle-deep in sachets of soup, loose potatoes, maps, cables, blankets, water bottles, bolt-cutters and maybe the head of an axe.

Edward O. Wilson, biologist and writer, describes these young protesters as 'the living world's immunological response.' Like antibodies they seem to appear out of nowhere, claiming to detect a dangerous virus, and place their bodies on the line. Tim Flannery says these protesters will proliferate in years to come. Like spiders, they are constantly spinning new webs. Some actions take an entire night to set up. If they are feeling spectacular and if conditions are right, they might knit all the machinery with cables and leave a single protester entangled in the middle. Nothing can be moved without the entire structure threatening to fall down. On scouting missions, the activists photograph and document each coupe, mapping out the waterways, recording species and keeping an eye on the harvesting procedure. Every now and then a white four-wheel-drive bearing the Forestry Tasmania and state government logos – the thylacine again – will pick up on their presence over the radio and chase them out of the back roads.

'We're the cannon fodder,' explains Wazza, a 25-year-old from Maitland, New South Wales. These post-apocalyptic 'ferals' or 'ratbags,' as many proudly call themselves, dot the emergency lanes of highways, knotted hair poking out from under their beanies, packs over their shoulders, thumbs out. Living as if the apocalypse has already happened, many have no future plans or engagements. 'Which is why we're useful – we understand urgency,' says Wazza. Their goals shift daily but all relate to one aim: securing protection for patches of nature, one jigsaw piece at a time. They survive by 'gleaning' or 'dumpster-diving,' recycling society's leftovers while angry supermarket managers pelt potatoes at their heads. 'That fucking hurt,' one gleaner tells me after copping a tuber to the jaw.

Short and stocky with blue eyes and cropped brown hair, Wazza has a picture of Ned Kelly tattooed on his leg. 'I got it when I was riding horses around South America. I was really homesick so I got this local tattooist to draw it on me. I told him the story of Ned Kelly. He loved it.' Fiercely intelligent – he provides me with a reading list and then a second list that counteracts the first – Wazza says he's always worked hard. 'It's just that some people don't see forest activism as working hard.' As a teenager, he worked at Hungry Jack's to save for university and is now on a scholarship, completing his honours thesis at the University of Tasmania. He has been part of direct environmental actions for six years and it doesn't take long to realise, whatever one's definition of work is, that he works hard at it.

Some of the activists are on welfare, but I don't think I've met anyone on the dole quite like this before, and that is probably what annoys their critics most: ferals don't even pretend to do what one is expected to with money. Wages, fortnightly welfare payments, savings, trust funds, even money raised from selling a Banksy stencil salvaged from a condemned London squat – all go towards the Florentine blockade. Instead of material comforts, saving for a home, paying bills, buying food or going out to dinner, their money is spent on cables, D-locks, tarps, abseiling equipment, sim cards, generators, tools and domain names. They can be found riding pushbikes along unsealed roads, hitching them over boom gates to dodge security guards, and setting up structures that can take days to dismantle, costing thousands of dollars in wasted work time.

'Lock-on' contraptions are built well in advance, as the longer cement is left to set, the stronger it is. These consist of concrete weights with piping threaded through the middle and a metal rod to hold on to. Old car wrecks are dug into logging roads, their wheels and ignitions removed. Using padlocks, handcuffs, thumb-cuffs, chains and climbing karabiners, the activists fasten them-selves to difficult to reach 'lock-on' points placed throughout the chasse. Bicycle D-locks can be placed around the throat of an activist, securing them to an obstacle and making it particularly hairy for them to be removed. Bits of metal pipe are welded together to make a V-shape; hands can be locked inside with climbing karabiners or padlocks. When locking on in small or difficult-to-reach places, such as beneath a truck or a machine, handcuffs are used. The activists won't tell me who holds the keys to these cuffs and padlocks. Perhaps it is the 'black wallaby,' an activist assigned to keep an eye on the 'bunny' (the arrestable) by hiding in the bush and keeping out of authority's reach.

They build tripods, wooden structures held up by cables on which activists can perch. Tree-sits are hung from the forest canopy, platforms with buckets for shitting and pissing dangling beneath them on ropes. A suitable spot is chosen by mapping out where a proposed logging road will likely go; the idea is to make it as difficult as possible for trucks to get into the forest. Once in their tree-sits, protesters are strung up by a series of cables, which will drop the sitter to the ground if cut. By law, logging cannot commence within a certain radius of these platforms. Demands to come down are studiously ignored by the activists. Wazza recalls watching the miniature ecosystem on the branch to the left of him instead. 'A ti-tree, about 30 centimetres tall, has taken root on the branch, 25 metres up in the canopy,' he wrote in his notebook. Sometimes the tree-sitters are smoked out; loggers make a campfire at the foot of the tree and burn all the plastic they can find. Search-and-rescue police shoot rubber darts trailing fishing line into the trees, trying to hook the line over a branch so they can thread a cable through and send a climber up. As soon as an arrow is shot, the race is on: the protester drops from his or her sit, abseils to the branch like a spinning spider and quickly cuts the fishing line before police can feed a rope through. 'Once a policeman is on the line, we don't cut,' one activist tells me.

In the Styx Valley, close to the Florentine, a 21-year-old activist spent fifty-one days halfway up a 75-metre tree. When it proved particularly difficult to get him down, search-and-rescue police eventually resorted to a helicopter, dropping two rescuers down on a cable to the platform.

'He was certainly surprised to see us,' says Inspector Brett Smith, who oversaw the operation. After his initial shock, the activist quickly helped the two policemen onto his sit and descended to the ground with them. The platform, which had been his home for a month and a half, was dismantled within three hours. The vigil tree soon spluttered bark as a chainsaw began at its base.

Early last year the SWST crew tried something new after they got an anonymous tip that the police were planning a bust at the

Florentine. 'There's something you should know,' said Ula when she greeted the inspector at the mouth of the blockade. 'There's a 5-metre tunnel beneath us and we've got someone in there.' The inspector stared at her, his hand motioning for the other officers to slow down. The activists had worked for months on the tunnel, shoring it up with wooden ribs like a mineshaft. It took input from a mining engineer, a 30-degree day and the police cutting off the ventilation pipes to coax the protester from the dark, wet hole – thirty-four hours later.

Before the Florentine blockade, mostly the same group of activists built an ark out of left-behind logs in the Weld. For fourteen months, the ship sat astride a logging road, blocking machinery and trucks. 'It was beautiful,' says Bridget. Photos show a pirate ship, a papier-mâché mermaid on its bow, seeming to sail out of the forest. Young activists from all over the world still come to Tasmania and ask to see it. 'We heard there is a ship in the forest, or some boat in the middle of the Weld,' they say – but the ark is long gone. A police bust cleared out the protesters, leaving only a couple of tree-sitters behind. One stayed up her tree for three days. She watched as loggers took photos on the deck of the ship with their girlfriends and wives, linking arms and singing sea shanties. Then they smashed it up, bulldozed it into piles and burnt it. A year later, when one of the activists met with Forestry Tasmania officers at the nearby Geeveston Timber Centre to discuss a moratorium on logging and protesting, he spied a photo of the ark on the office wall, beautifully framed. Observing his surprise, the forester explained: 'Well, it was pretty impressive.'

Then there was the Angel. In 2007, the activists decided to block the only two ways into the Weld valley. One activist was harnessed beneath a bridge, making it too dangerous to cross; further in, another was locked onto a boom gate across a logging road; and atop a giant wooden tripod, blocking the second entrance to the forest, was the Weld Angel. Allana Beltran, twenty-two years old at the time, had made her wings out of cockatoo feathers collected from roadkill and hitched them onto her back at a slightly lopsided angle, so that when she moved she looked like a bird adjusting its feathers. Her face was painted white and she had tiny white earphones tucked into her ears, an iPod hidden in the cloth swathed around her. It was dawn when the first trucks braked in front of her. With the golden light streaming through the khaki canopy, it is said some of the workers were speechless, climbing out of their trucks to stare up at her. For as long as her iPod battery lasted, the angel listened to Tibetan monk music while police ordered her down through their megaphones. Images of the Weld Angel soon proliferated on the web, even turning up at a New York fundraiser for African rainforests attended by a clutch of celebrities.

But neither the ark nor the angel managed to protect the forests. The Weld had been classified as a 'deferred forest area,' which David Llewellyn, a minister in the state Labor government, had said would not be logged. At the last minute, however, the Weld was dropped from a collection of new reserves announced after the 2004 federal election. Two years later loggers slowly drove a single rider through Hobart. The tree was 200 to 300 years old, its girth 8 metres. On its base, the workers had spray-painted a message to the then leader of the state Greens: 'To Peg Putt with love from the Weld.' Other trees hauled through Hobart bore the message 'Joy to the Weld.'

No one knows why the Weld was sacrificed, although some suspect it had to do with an action by these particular activists in the final hour of decision-making. When I ask Wazza about this theory, he admits cautiously that there might have been a screw-up.

'We've had some stuff-ups. Once we floated all our gear on a zodiac [dinghy] down the Weld River and set up an obstacle over-night. In the morning we realised we'd built the blockade in the wrong place.' He laughs. 'But there was an incident with Senator Eric Abetz that might have seriously backlashed against us. One of our crew was supposed to handcuff himself to a bit of signage at the press conference, but he got confused by the others' hand signals and handcuffed himself to Abetz instead.' He sucks in his breath sharply. 'Apparently Abetz flipped out and people say the Weld, which was our campaign, got dropped from the reserves soon after. But no one knows for sure,' he quickly adds.

The activists tread a fine line between drawing attention to threatened areas and provoking resentment that can ultimately backfire against the forest. Have ratbag actions taken their cause backwards rather than forwards? Today the Weld is still being logged. Boom gates have been erected outside the state forest and visitors must leave a fifty-dollar deposit to borrow the key, which must be returned within twenty-four hours. Most of the activists are now committed to the Florentine blockade, but there are stragglers, deeply attached to the Weld, who have held on, still lugging chains and equipment through the Huon Valley scrub to set up roadblocks and hold off the ever-growing number of trucks, dozers and men.

All of this takes place away from the main road. The pull of bitumen steers most people away from the strange gatherings under the trees; boom gates keep the two worlds firmly apart. One of the few small windows between them is the Hobart courts. A typical day's proceedings tends to run something like this: drink driving, assault, assault outside a nightclub, trespass into the Florentine forest, girl punches another girl in the face in a shopping centre, shoplifting five CDs, refusing to leave a pub, obstructing police work in the Styx Valley, drink driving, drink driving again, assault, road rage, trespass into the Weld Valley. On the day I sit in, a judge tells a woman up for drink driving that her blood-alcohol reading means it would have taken her sixteen hours to be completely sober and eleven hours to even consider driving. The courts are dotted with activists quietly waiting their turn to be sentenced. When they do stand, dreadlocks pulled into ponytails, sleeves pushed down over tattoos and septum piercings tucked discreetly into nostrils, others in the room roll their eyes. It is a curious sight: a surfacing of passion, of a commitment to a higher ideal, in a court more used to bored insolence.

CAMP FLOZZA

To the uninformed eye, Camp Florentine looks like a shit heap. Which is how it looks to me. The torched cars are still lingering like a hangover next to the road. Rectangles of sunlight spill through the axe wounds onto melted seats and burnt calico shopping bags. The stink of rubber catches on the wind. Poking out onto the main road, the camp is built on top of an unseemly burp of tree stumps and gravel – the beginnings of a logging road. Under a tarp two fellas, one scruffy and shoeless, the other tall, lean and classically beautiful, stoke a small fire. A staffy lies between them, his pink dick unfurling in and out like a lipstick. Brochures with information about the area's walking tracks are weighed down with rocks and a handwritten sign asks passing cars to 'Stop for a cup of tea.' I ask how many cups of tea they've made for tourists in the past decade. About two, the taller man replies, grinning. His name is Ali and he is Pakistani; the other man, bearded and weary, is known simply as 'No Shoes.'

'Welcome to Coupe 44A,' Bridget says wryly, gesturing up the dirt road. A sign reading 'Camp Flozza' pokes out of a mound of ferns and debris that has been bulldozed up both sides of the road like a permanent muddied wave. Of the planned 10.8 kilometres of road, Forestry Tasmania has so far managed only 300 metres thanks to this mob.

Following Bridget, we walk up towards a small hut nestled between two tall trees known as the 'guardian trees.' I dodge a cobweb of ropes and cables woven across the road. In one of the guardian trees, a tree-sit about the size of a single bed hangs from the canopy. The hut is made out of salvaged corrugated tin and wood. Underneath it, two old cars have been star-picketed into concrete barrels and fastened to the ground. Activists are rostered to sleep in them each night, ready to lock on at any sign of a bust. It is like the aftermath of a bush rave. There are hand-painted banners with references to mother earth, fluoro string tied randomly around dead stumps and assorted hippy imagery, all of which makes me want to cover up my chakras and run. And in a sense I've been doing exactly that these past seven years. Beneath my clothes I sport various tattoos of totemic animals, Celtic symbols and hippy spirals. When I was seventeen I was so stoned getting one tattoo I didn't notice until months later that the symbol of Christianity, the outline of a fish, resided in the middle of it. It's hard to know if this former skin of mine will make me more understanding of or harder on this ratbag crew. I do know that if I were still a hippy, I'd think this situation 'serendipitous.'

A couple of young activists sit by a fire with a pot of water eternally on the boil. It's cold, and every now and then one stands up and stamps some warmth back into his or her body. The rickety kitchen looks like a tourist bar on a Bali beach and once featured on *Going Bush*, a bizarre TV show produced by Forestry Tasmania. Every Sunday night, the show's presenters would interview 'guests' such as Bob Gordon, the managing director of Forestry Tasmania. One week, they decided to pay a visit to the blockade.

'They totally surprised us,' Wazza tells me later. 'Just showed up with their cameras and presenters and asked for a tour.' It was one of those do-or-die public-relations moments, and he agreed to show them around. When I finally watch it, I find that Wazza turned out to be a match for the self-described 'inimitable and laconic' presenters. Gesturing to the kitchen, he tells the camera that this is where the activists enjoy 'caviar, roast duck, and all the top-end Tassie cold-clime wines.'

'They kept stealing all my lines,' he recalls, laughing. 'I'd say something funny and they'd stop the cameras, then start filming again and say what I said.' Tough to respond to your own funny one-liners with even funnier one-liners – but he managed to pull it off, coming across as easy-going. Rather than arguing the activists' case, Wazza stuck to talking about everyday life on the blockade. 'It wasn't hard to see what they were doing, so we just played along and rode it out. There was only one stuff-up at the end.'

After touring the camp, the presenters spied a guitar and asked if anyone would play something for the camera.

'Me and Uls stood behind the presenters, waving our arms and stabbing at our throats mouthing "no," but one guy took the bait,' Waz says. A hippy playing a bad folk song on an out-of-tune guitar was the last image the SWST crew wanted to convey. As the long-haired kid struck up a tune, they started to bang tools in the background to screw up the recording, but the cameraman kept filming. 'Of course, the guitar was the opening scene,' Waz shrugs.

Standing in the kitchen, surrounded by pots suspended from nails hammered into branches, I spy boxes of food. Some have been donated and others dumpstered. Hungry, I peek into them. There are loaves of homemade bread, with notes attached listing the ingredients in an older person's almost unintelligible scrawl. I make out 'rosemary,' 'olives,' 'pumpkin,' 'dates,' 'gluten-free' and so on. In another box I spot a container of gourmet vanilla-bean yoghurt past its use-by date. I peel its lid off and have a sniff. Satisfied, I borrow a spoon from the cutlery jar.

Wandering away from the campfire, I find myself in a clearing where knocked-over ferns have been shakily propped up. They look silly and sad. Miranda Gibson, the young woman from the video footage that prompted me to book this trip, joins me. She has brought a spoon and we share the tub of yoghurt. One of the camp's core climbers, she is softly spoken and unassuming. All the tree-sitters are referred to as 'possums' to keep police from catching their names, but Miranda's gentle nature is the most possum-like. Wearing a baggy T-shirt and baggy pants, she is small and pretty, with straight brown hair hanging shyly over her eyes. She grew up in Brisbane, where her dad is active in fighting for workers' rights and her mother is a feminist. A special-education teacher's aide, she has spent most of the past four years at the blockade. 'I used to go into [the nearby town of] Maydena and do aerobics with some of the women there. I even had some baking days and cake stalls with them. But I won't go there anymore.' Since being caught in the attack on the car, Miranda has become more cautious. When the camp was torched, she watched from her tree-sit as men stomped around with axes, yelling at the protesters to come out. 'I prefer the trees. It's safer up there.' Most days and nights she climbs the 25 metres to her tree-sit, rotating shifts with another climber. In her long stints in the canopy, she reads novels, writes letters and collates wildlife reports about the plants and creatures of the Florentine.

I also meet Cess, an older woman sitting cross-legged and painting banners; Rosie, an eighteen-year-old from Adelaide who does a pretty mediocre job of juggling; Dylan, whose toe is infected and looks like it's going to fall off; Holly, an activist from the UK with long ginger plaits and her Australian husband, 'Mad Dog';

Wendy, a cynical older hippy with grey dreads; and Petal, a crazy Kiwi who found these kindred spirits while hitchhiking: 'This crappy car stopped for me and I had this plastic pirate's sword under my jacket and I pulled it out as I jumped in and said, "Arggggh, me lovely!" They didn't blink and I ended up going to a party with them that night.'

There are others, some aloof, others shy. At first I assume they all know each other, but in time I realise that some don't know the first thing about their companions. There is an odd lack of curiosity in the camp. People float in and out, asking few questions of one another, as if the past is erased and this, what they are now, is all that matters. I find this depressing. It's as though they see personal history as mere gossip, irrelevant to the cause. The men and women who started the blockade could walk in and no one would know who they were; their attachment to the place would be seen as no more important than that of someone who had been here for a week. When I say things like, 'You know, so and so, the ballet dancer,' or, 'I can't remember his name, but he gave a me a lift, you know, the guy whose folks are missionaries,' I'm met with blank looks. 'Yeah, a lot of people come here carrying a lot of baggage,' says one activist, 'and we should be more there for each other, but often there's no time.'

On my first night at the blockade, I sit quietly as hippy mush is ladled into bowls. As we eat, the blockaders work out the next day's roles – who's on night watch, who's on morning watch, who will be sleeping in the cars, who'll be up in the tree-sits – and agree on new signals for cops, thugs and uncertainties. After dinner, some dump their bowls near the sink and pull away from the fire, swallowed up by the night. Not much is said and I'm not sure if it's my presence, people weighing me up. I say goodnight and take off with my sleeping bag. Miranda does too. She knows the nooks and footholds of her tree and can climb up and down in the pitch black, reading the bark like Braille. She is used to the midnight visitations of creatures, listening for the swish of sugar gliders, the gloomy foghorn of boobook owls. But I feel awkward with the forest. I end up wriggling into my sleeping bag in the makeshift hut, zipping it all the way over my head and breathing hot air into my synthetic sack before unzipping and gasping in the cold forest night. I shut my eyes tight, curl up like an armadillo and will myself to sleep.

*

It was Prue Barratt, a petite, foul-mouthed local, and Stuart Primrose, a clean-cut self-employed arborist, who started the blockade in the Upper Florentine. Originally the camp was made up of locals and called itself the Derwent Forest Alliance. Stuart stayed in a tree-sit, climbing up and down for three months, while Prue set up camp below, sleeping in her hand-painted panel van. Other members of the group brought supplies but the usual excuses – family, job, money, life – kept stopping them from joining the camp or taking over for a few nights. After discovering a dilapidated hut belonging to the famed tiger trapper Elias Churchill, the group received a $6000 grant to restore it. But when they had done the basics – re-laying the foundations and so forth – Forestry Tasmania took over the site, claiming it as their own, and finished the restoration.

'That was weird,' says Stuart. 'Obviously they wanted the kudos of restoring it and not to let some protest group draw attention to its historical value, but why fix it up if you're planning to log all around it?'

For almost five months the two entertained visitors, taking them for walks and telling them about the area. About a kilometre from the blockade there is a cave containing the oldest human remains ever found in Tasmania, carbon dated 10,000 years old, Stuart tells me. Underneath the Florentine there is a network of unexplored caves where the river dips in and out, weaving through darkness before dashing up for air. 'People have found the remains of mega-fauna around here as well,' he says. 'Enormous wombats, that kind of thing.'

When a storm was coming, Stuart said that from his tree-sit it sounded like a train roaring down the valley, and in the winter he had to push snow off his platform. But things became difficult – numbers dwindled and the blockade wasn't strong enough to resist a bust. 'We needed more people, and so the ferals moved in,' says Stuart. Intentionally or not, the ferals took over.

Three years later, Stuart is still undecided about them. 'I mean, it seems pretty hypocritical for me to criticise them when they're the ones still there. No one else is doing it. But …' He trails off, unable to put his finger on his feelings. Prue, however, has no problem articulating it. 'We were good. I kept the camp clean, we kept clean, ate well, spoke well and didn't scare visitors off. Only problem was there were only two of us. Now there are heaps of them, but they won't change. If something doesn't work in a campaign, then you change it. Everyone knows that. But they won't change. They expect everyone to change but themselves. Yeah, in an ideal world you wouldn't be judged on how you look, but in an ideal world, would a rainforest be logged?'

For Prue, the shine started to wear off her fellow blockaders after her first lock on. 'Hey, Prue,' a voice whistled as she lay awkwardly on an old mattress, her arm tunnelled through the foam and locked into cement. 'Smile for the camera so I can show our mates what you do now.' Recognising the voice of her ex-boyfriend

40

from high school, she jerked her head up and he took a photo on his phone before walking off to join his workmates in the timber crew. Initially the cops ran right past her, blind in the excitement of the bust. 'They must have thought I was just lazing around, having a nap amongst all the chaos. So they're all racing around, looking at the tree-sits, calling up to sitters on their megaphones and this Forestry guy recognises me on the ground and he says loudly, "Prue, what does your grandfather think about you doing this?" You could have heard a pin drop. Everyone was like, what the fuck – who is she?'

Prue says she had gone over everything in her head before her first lock on: why she was there, how the police would go about removing her and how long it would take. 'But not once did I prepare for what happened. I was really bloody naïve.' Wearing a hardhat, she pointed out to police where her arm was, where they'd have to dig to get her out and then obediently turned her face away. 'When the crowbar first scraped past my arm, I thought he must've slipped. I cried out a little to let him know how close he got. Next thing, the crowbar was actually being slammed into my arm. It was fucking agony. My arm started to swell and fill up the pipe.' An inspector was overseeing the operation, but Prue believes he made no attempt to intervene. If anything, she thinks he urged the second hit. 'I heard some muttering, I think another policeman took over. I'm not certain, though. After the crowbar hit a second time, I locked off.' Prue says she lost her mind a little after the incident. 'I was like, shit – I've no protection. I know it was childish to think that I ever did have protection. I mean, of course police are not going to be any better than other people just because they're cops. But it scared the hell out of me. I guess I just woke up to a reality that two-thirds of the rest of the world already has to live in.

Worse, I knew no one was going to believe me. I'd turned into one of those mentally ill conspiracy-theory greenies that I hate.'

But Prue didn't just look at police in a new way – she saw the blockaders in a different light, too. 'I saw that a lot of them were getting off on it, the thrill of being a rebel. I think it's negligent they don't clean up, negligent to the campaign and to each other. Instead of wearing burlap sacks and not washing, they could use their looks and intelligence to lessen the chances of the rest of us getting beat up. But their subculture is more important than the trees or each other.' Prue describes holding endangered-species days at the forest and activists showing up with their dogs. 'Or like when I'm talking to a truck driver and I hear this "tinkle tinkle" behind us and it's Harvey with bells in his dreads and on his toes. I mean, what the fuck? What truck driver is going to listen to some boy in hippy rags and bells?

'The whole dumpster-diving thing, being "freegan." I mean, it's cool for a moment, but it's also half-arsed. Their whole ideology is half-arsed. Get the media out there but then look shit for the cameras – I mean, what's the point? Why don't they complete the circle, get to the solution? Instead of eating from the bin, pay extra for organic produce.'

'But if you say that to them,' I venture, 'they'll pack up and leave.' I sit up straight and in a school-marmish voice, as if reading from a report card, add, 'They've got so much potential.'

Prue laughs. 'Doesn't everyone?' she says.

*

'It would definitely help our cause if we didn't look like ratbags,' admits Wazza. 'The Greens had the right look in the '80s. They looked like hikers and librarians. They had a collection of conservative second-hand clothes for protesters to wear. They called it the "camouflage cupboard." But at the same time – we're people. It's all very well for people to say we should ride our bikes to blockades, be strict vegan, not smoke or drink or have kids and dogs with us, but that is just unrealistic. Yeah, there are some kids who just use our blockade as an excuse to do nothing or play out some anarchist fantasy, but they're not the core crew.' But Waz admits he is starting to step back, get some perspective on things. 'I'm thinking more about the consequences of what we do.' In 2009 he cut off his dreads and bought a pinstriped suit. He started to meet with people in Canberra. 'But underneath it, this is where I come from. I'm a rat-bag, a fez.' And he means it literally, rolling up his shirtsleeves – there are inky reminders etched into his skin.

'We can't force people to be different to who they are,' another woman cuts in.

'But what if it's to the campaign's detriment?' I ask.

She shrugs. 'People will say what they want about us no matter what we do.'

I shake my head. In a world of sound-bites and five-second grabs of footage, I can't help thinking that if they're going to go to the effort of living out here and getting the media to visit, they could smarten up. A woman with blue eyes and perfect teeth reads my mind. She holds out one of her sandy-coloured dreadlocks to me; it's the length of her arm. 'But we live in the forest at a blockade,' she says, gesturing at her muddy black hooded jumper with patches safety-pinned to it. 'This is what happens.'

A few months later, I attend a rally at the blockade. Ali has shed his hoodie and muddy pants and put on a suit. It's silky, too suave a suit to be wearing in the forest, but he's decided he has to look smart. I watch him pick his way through the crowd, asking people to move for cars to get through, trying to keep the peace as people push up against a wall of police.

Later in the day I see him again. People are now perched around orange bulldozers and singing, but he's rattled, his face pale. 'I'm never wearing a suit again,' he says. He has been elbowed in the guts by feral activists, who hissed 'Forestry scum' as they pushed past him. And one of the sergeants, pulling him aside, said quietly in his ear, 'I know what you are under that suit, Ali. You're scum, you can't hide from me under a suit. You can't dress up scum.'

SPLITTING HAIRS

B ridget likes to carry a stick or a stalk of long grass and swing it like a whip as she walks. She does it dreamily as if not aware of it, but twice she stops abruptly and we crash into the back of her. Over her shoulder Miranda and I see tiger snakes slowly unfurl and sliver off the path. We are walking to the Lookout, following an old supply track that the osmiridium miners used. On one side of us is protected World Heritage-listed forest; on the other is the Upper Florentine blockade. Birds whip across the path, indifferent to such boundaries. Miranda crumples a leaf and holds it under my nose to smell its sweet scent. 'Chew it if you want,' she says. 'It's sassafras. The Aborigines reckon it's a pick-me-up.' We stop at a large tree the activists have named the Twisted Sister because it coils into the sky like a whip of liquorice. It is some 80 metres tall and its base is the size of a small car.

These tall eucalypt trees – the stringybark, messmate, alpine ash, swamp gum, and mountain ash – are among the tallest trees in the world. The regnans, literally translating as the 'King of the Euca-lypts' and found only in Victoria and Tasmania, is officially the world's tallest flowering plant, sometimes splintering the sky at over 100 metres. In the 1860s Australians took particular pleasure in declaring the height of the regnans greater than America's beloved redwoods. Instead of increasing their standing value, this competitiveness got people into such a tizz that the giant trees were cut down to be measured. Knowing where a giant tree resided became a kind of power, with debates played out in newspapers over localities and measurements, and American patriots arriving to investigate.

Today many conservationists like to call the regnans the 'light-houses' of the forest; they extend high above the canopy as if to watch over a green sea. In their old age, 300 to 600 years old, they are the perfect site for an eagle's nest – the bare grey branches jut out like antlers and the enormous birds can sit on their stash of twigs and eggs without losing sight of the vast stretch of forest. Environmental campaigns seek to protect the regnans because of their height. 'It's a media stunt,' says Miranda, frustrated. 'You have pamphlets saying things like "sixty people can fit inside the hollow of this tree," but something shouldn't have to be big to be protected. People come here and want to know how old each tree is. I tell them the tree is between 200 to 600 years old but this forest, it is thousands of years old.'

In response to such campaigns, Forestry Tasmania announced a policy to protect 'giant' trees, only to adjust the definition of giant twice when it realised logging was going to be grossly inhibited. At the moment, a giant tree must be over 85 metres tall or 280 cubic metres in volume to be protected. But this is an impossibly static requirement for plants, says Miranda. Unlike Australia's other favourite giant things, such as the Big Pineapple and the Big Banana, these are not made of fibreglass. 'As trees get older, they shrink. Like people, I suppose,' she says. 'They drop their crown, which also means a giant can become a non-giant if and when it drops its crown. And only a eucalypt can get to that height, so the biggest myrtle, even if it is 500 years old and the biggest of its species, isn't counted.'

At another tree, Miranda points at leaves that look like celery shoots. 'Celery-top pine,' she says. The tree is spindly. I'm able to get two hands around its trunk, but give it another 400 years and it will pack on enough wood to make the perfect boat plank. Water-resistant and lightweight, celery-top pine is much sought-after by boat builders. 'But this coupe will be logged on a thirty-five to ninety-year rotation and this skinny tree will be too immature for anything.' Miranda insists the blockade is not naive. 'I'm happy for a tree to be cut down, but logging here is done in such a rushed and greedy manner. They trash everything for woodchips, putting aside a few sawlogs if they can be bothered and burn the rest to replant a commercially viable forest and then call it regrowth. But it's not regrowth, it will be nothing like what is here now.'

Sowing seeds from a helicopter or a fixed-wing aircraft, Forestry Tasmania claims to use a 'specific seed recipe' that has been gathered from the local area. That may be true, say these activists, but creating a hot ash bed before showering the flattened coupe means a specific type of tree is being advantaged. Two types of tall eucalyptus trees – the *Eucalyptus delegatensis* (gum-topped stringybark, alpine ash) and the *Eucalyptus regnan* (mountain ash, swamp gum) – regenerate best from seed and need heat to be released from their gumnuts, unlike other eucalyptus species which tend to re-shoot after a fire. They also need a lot of light – hence the best ground for them is an empty hot coupe. As for the blackwood, myrtle, celery-top pine, laurel, young Huon pine, wattle, ti-tree, leatherwood, banksias, musk and other eucalypt trees, as well as the mosses, fungi, bracken and ferns, all of which SWST have recorded in the designated coupes, most will struggle to return in time, if ever, for a second harvest.

Thanks to the tall eucalypts' ability to stand straight in a country where most native trees tend to flail about, sinewy branches and wrinkled armpits twisting like the limbs of a contortionist, it is these trees that have been in high demand since the beginning of Australia's timber industry. It wasn't long before people started to refer to them, along with other types such as *Eucalyptus obliqua* (stringybark), as 'industrial plants.' More like the good 'up and down' European tree than an experimental study in sunlight, they became even more sought-after when it was discovered they were also the best native trees for pulp and paper manufacturing.

When in 1935 a distinctly piss-coloured issue of the *Mercury* newspaper was printed on local paper, their fates seemed sealed. Sold as 'Tasmanian Oak,' a blanket term applied to various types, such trees are now mostly used in pulp, flooring, veneer and structural building products. They are cut in their commercial prime, at about 100 years old, but the timber industry's idea of 'prime' is often only the beginning of such a tree's role in a dynamic ecology.

About a quarter of Tasmania's native birds and a third of its animals use the old tall eucalypt trees as homes. In forests like the Upper Florentine, many of the biggest and oldest trees are simply in the way of logging. Too dangerous to work around, they are cut down or loaded up with dynamite, loggers triggering the explosion when they are a kilometre or so down the road. Timber workers are known to call out amicably to activists, 'Don't worry, love, it's rotten up the inside!' to highlight their uselessness. 'Exactly,' Bridget says to me. 'So don't chop it down. It takes about 150 years for a regnan to start forming hollows,' she explains. Under one such tree, the girls point out various places where hollows could be forming. 'It's like a block of flats.'

*

As the three of us walk further along the track, the canopy starts to fall away until we are walking through short scrub, then rusty button grass. A long blue sky stretches out and yellow prairie light falls on tiny spiders lassoing reeds together with web. When we reach a small lookout, we climb onto it to look at the outlines of mountains pressed like bruises on the horizon. We rest on the warm wooden slats of the platform, splaying our arms out in the sun. I think about my road map and how the Upper Florentine Valley looks like a speck in comparison to the huge swathe of protected World Heritage land nestled alongside us, which starts at the waist of the island, goes west and then nooks down into its southern valleys.

It is a significant proportion of breathtakingly wild country. There are dolerite cliffs, formed by fountains of volcanic lava, which look like enormous cement French fries, and tiny glacial lakes, where primitive shrimp spawn, breed and die in a day, as they have done since the ice age. On the west coast, there is a tree that takes up an entire hectare. Estimated to be over 10,000 years old and one of the world's oldest living organisms, this Huon pine has repeatedly taken root as centuries of snow weighing on its branches slowly anchored them to the ground. Beneath rivers, basalt eruptions have revealed trifle-like layers of Jurassic forest and petrified ferns. And all this, approximately 1.4 million hectares, has been protected, albeit begrudgingly.

'Don't you think that's a lot of land already protected?' I suggest cautiously to my guides. Miranda nods and takes a deep breath, as if preparing for a discussion she's had a hundred times over.

'It's definitely a lot of land.' She unfolds the maps I asked her to bring. Laying them on the platform, she uses my elbows and Bridget's feet for paperweights. Pointing to the vegetation map, Miranda traces her finger down along the World Heritage boundary, emphasising where the line shirks dark green patches, forests with tall eucalypts in them. 'What would loggers want with mostly button-grass plains, rocks and mountains?' she asks. As she traces her finger down along the 'unlucky side' of the eastern boundary of the World Heritage area, I read out the names of the forests – 'the Styx Valley, Upper Florentine, Middle Huon, the Weld and Upper Derwent …'

'… the Arve, Denison River, the Wedge,' continues Bridget.

All have been identified by experts, including the World Heritage Committee, as being of equal value and importance to the large reserve they're nestled alongside, but when it came to drawing the boundary, many of these forests were left outside.

'But what about the east?' I ask. 'It's like the eastern half of the island has been ignored.'

Miranda nods. 'It's a "reserve this and smash that" mentality,' she says, brushing her hand across the north-east. 'There has been a massive clearing of forests over there to start plantations. But apparently that's okay because of this reserve over here.'

Over the past thirty years in Tasmania there has been a sausage string of forestry inquiries, reports, impact statements, court cases, legislation and subsequent amendments, Royal Commissions and agreements, including the Regional Forest Agreement in 1997 and the Tasmanian Community Forest Agreement in 2005, both struck around federal election times. All claimed to be the last of their kind, but each new piece of legislation seems to have further entrenched the conflict and obscured the issues. These reports often rely on sweeping generalisations and simple statistics: '45 per cent' of Tasmania's forests are protected, '95 per cent' of 'high quality' wilderness is in reserves, '79 per cent' of all old-growth forest is safe from logging. These statistics are like a swarm of small truths that move too quickly to be properly scrutinised. And the problem with small truths, I've often found, is that they don't necessarily make a big truth.

Just under half of Tasmania is covered with forest – 3.3 million hectares to be exact. As the island's former premier Paul Lennon once said, that's 'a lot of fucking trees.' But trees, I'm soon to realise, don't necessarily equal forests – after all, the state government includes plantations and 'regrowth' forests: forests felled and then re-grown to be more 'wood productive.' In 1997, in preparation for the Regional Forest Agreement, a group of scientists mapped out fifty-one types of forest with a view to developing criteria for managing them wisely. It's said the agreement – which outlined the fate of Tasmania's forests for the next twenty years – was based on these findings, but Professor Jamie Kirkpatrick, a respected botanist and member of the scientific panel responsible for designing the criteria, said their work was a shadow of its former self by the time the document was signed off. 'The RFA got perverted in its later stages into a political process which had nothing to do with science,' he said.

In a paper Kirkpatrick wrote after working on the RFA, he described the editing process the criteria underwent.

> While much of the wording of the original criteria document remained intact, additions and deletions were made which apparently were designed to avoid any necessity to conform to any concrete targets ... The words 'where practicable and possible' were inserted to modify the strength of a large proportion of the criteria.

Complicating matters further is the fact that 30 per cent of the 3.3 million hectares of forest is found on private land, where it is largely unprotected. The Regional Forest Agreement included an initiative to build up substantial reserves of the various forest types on private land, with the goal of protecting at least 100,000 hectares of private forests; the federal government gave the state $30 million to kick-start an acquisition process and fund steward- ship incentives for landowners. At the same time, however, the federal government was also pushing for a mass roll-out, largely unregulated, of plantations, and the state government was setting up 'private timber reserves,' allowing landowners to log their native forests or start up plantations (or both). If approved under the Forestry Act, these new 'reserves' will be exempt from standard rural planning laws, local government powers, cultural heritage and environmental management laws. Conservation groups esti- mate that about 100,000 hectares of native forest have been cleared to make way for tree farms since 1998. In just over ten years, the state's plantation estate has leapt from 64,000 to almost 300,000 hectares. Meanwhile, the initiative to protect forests on private land fell more than 60,000 hectares below target and was wound up in 2006.

It's difficult to picture a hectare. On average 15,000 hectares of forest are logged each year, 12,000 of which are in state forest and 2000 of which are defined as 'old-growth.' Local street spruikers for environment groups say it's the equivalent of '10,000 football fields a year' but on further prodding, many cannot say if this is your average local footy oval or the Melbourne Cricket Ground. It turns out the MCG's oval is just over 1.5 hectares. The Opera House covers an area of 1.8 hectares. Melbourne's Federation Square is 3.2 hectares. Still, it all sounds like a pamphlet – unless, that is, you live in Tasmania. Here it doesn't have to be imagined. On the streets of Hobart, the nation's second-oldest city, scenes from the forest regularly visit its residents. Fifty fully loaded log trucks can pass through the city in a single day.

When I try to find the '79 per cent of protected old-growth forest' on the vegetation map, Miranda laughs. 'You'll be more confused when that statistic starts growing.' She says that the figure will continue to grow no matter how much forest is logged or burnt. 'It's a percentage of a diminishing pie, not of the original forest cover. And the smaller the pie gets, the bigger the portion of old-growth will appear to be.' This is the kind of maths that hurts my brain. It is like negative space where you have to try and reverse your depth of vision. Nevertheless, I need to split hairs. I need to pan out from this lookout and blockade to get a sense of what I'm dealing with. So here goes. Of the island's 3.3 million hectares of forest, 40 per cent is protected, but about 10 to 15 per cent of this is 'informally' reserved. For clarity's sake, let's say 1 million hectares of forest is properly protected and 400,000 hectares is informally reserved, leaving unprotected just under 2 million hectares.

Or, put another way, before European settlement Tasmania had an estimated 4.8 million hectares of native forest. According to Forestry Tasmania, over 1.4 million hectares of these forests are now protected. There are two ways to interpret this. From a conservationist's perspective, that means 30 per cent of Tasmania's original forest is protected from logging and 70 per cent has either been destroyed or is still available for logging. Whereas pro-logging types will point out that that's a large portion of the current island 'locked away.' Dissecting the island's forest estate is a nightmare. Everyone wants to start at a different place. Are we talking about forest cover as of yesterday, or forest cover as of 200 years ago?

Many reserves, activists also warn me, are dispensable, especially when it comes to an area that has high rainfall, good soil, tall trees and mining prospects. 'Informal' reserves run by Forestry Tasmania can have their protected status revoked with the stroke of an administrator's pen, while even 'formal' reserves are seen by some as an exercise in public relations – conservationists claim that much of the so-called 'reserved' forest is unsuitable for logging anyway, or should not have been included in logging coupes in the first place. In 1997, for example, when the Regional Forest Agreement was announced, much fanfare was made about 293,300 hectares of forest being moved into reservations. At the same time, with zero fanfare, crown land, notably the Tarkine rainforest corridor, was quietly transferred to Forestry Tasmania, the state's forestry-industry body and the largest supplier of timber to companies such as Gunns. Eight years later, the same tract of forest was included in the 2005 'reservation package' – the rainforest is shuffled back and forth, in and out of timber production like a political pawn.

The statistics become even more hellish when one tries to figure out what type of forest and how much is currently reserved – and this is the clincher, the fracture line. It turns out that the generous protection of forests has not been extended to the types of forest logging companies want: seven forest types have less than 15 per cent of their current extent in reserves. Six of these are dry eucalypt communities and one is a tall wet eucalypt community – parts of the Upper Florentine, for example. Of these lesser-represented forests, the extent of old-growth left is receding dramatically. Then there is the murky issue of what meets 'old-growth' classification, reinforced by the fact that the tallest living regnan in the world, named Centurion, standing at 99.6 metres tall, was found in a patch of 'regrowth' forest.

Of the tall wet eucalyptus old-growth forests, only 191,100 hectares are left and less than half of this is protected. In the regnans' case, 12,000 hectares of old-growth remain and 5000 hectares are formally protected. But these figures are already out of date, the forests last having been comprehensively re-evaluated for the Tasmanian Community Forestry Agreement in 2005. Considering that between 1996 and 2006 over 10,000 hectares of tall eucalypt old-growth forest were cleared, mostly through logging, today's numbers may be much lower.

Dividing forests into 'types,' however, gives an unlikely impression of a neat landscape, one that can be tidily divided up and 'harvested' accordingly. In reality a single area can include dry forest, slope into wet eucalyptus and then into a gully of rainforest. Even the term 'old-growth' can be misleading, suggesting a bunch of old and rickety trees. In 2007, when polls showed that most Australians were opposed to logging old-growth forests, Bob Gordon of Forestry Tasmania told the *Australian* he thought the public were being sentimental. 'Old-growth doesn't live forever,' he pointed out. 'Trees die.' For the most part, however, 'old-growth' forests contain trees of all ages, with new and old plants falling and rising at the same time, like any other living community. The idea that a circle can be drawn around one type of timber is increasingly out of date.

It is not until I surround myself with billowing maps showing tenure, vegetation, changing boundaries, contours, that I begin to glean the 200-year conflict over this island's natural resources. At the library there are maps that look like they've been drawn with squid ink – the faded lines tapering into the unknown and marking with squiggles the possibility of mountains. Lines are constantly drawn, erased, shifted a little to the left and redrawn. Like a missing jigsaw piece, there's a strange hole in the Hartz Mountains National Park – 'The government swapped it for the Precipitous Bluff forests ages ago,' Miranda had said.

No one map of Tasmania seems to be the same. On maps provided by the timber industry, the protected areas are colour-coded red. On maps provided by activists and campaigners, old-growth forests to be logged are red. On some tourist maps, towns and natural attractions have been struck off without warning. Some locals say the Hellyer Gorge was removed when surrounding myrtle rainforest was 'over-cut' and carelessly logged all the way to the edge of the highway, but others say the 'black ice' made it a dangerous touring route. The townships of Preolenna and Meunna were taken off the Regional Touring Route map not long after vast swathes of the former farming country was replanted with plantations, the absentee ownership of these tree crops creating ghost towns.

Standing on a lookout with the maps spread out around us, I can imagine how easily deals might be done in boardrooms, where wilderness is reduced to abstract numbers of hectares and its fate sealed with a handshake. The risk, I realise as Bridget chews her walking whip beside me, maps fluttering around us, is in *seeing* a place. Which is perhaps why in the early '80s the prime minister, Bob Hawke, vehemently rejected an invitation to walk through the state's forests with the Greens leader Bob Brown. 'I've seen what happened to Richardson,' he said, referring to his environment minister, Graham Richardson, who became an advocate for forests after a day spent in the wilderness with Brown. 'I'm not going near those forests with you.'

I wonder if I'll ever be able to refold these maps as perfectly as I found them. I imagine making a chronological flick book with them instead. An animated version of this island sped up over time where the wilderness is eaten up slowly as if a woodborer is making its way across the landscape, and every now and then, the wild takes a bite back.

MONKEY-WRENCHING

In December 2008, seventy police converged on Devonport Airport in northern Tasmania. Their code name was Western Approach. The tinny voice coming from the hearing aids in the officers' ears urged calm. Armed forest activists had hijacked an aircraft, threatening to fly it and its passengers into the Wesley Vale pulp mill. Their confidence faltering, the activists transferred their hostages to the terminal and started making demands. Slowly, creeping like crabs along the periphery, the special police moved onto the tarmac and surrounded the airport. The activists let the hostages go, one by one at first and then in groups of two and three. Eventually the hostage-takers emerged from the airport. Still posing a hypothetical threat, they were shot.

The bullets weren't real, this was a drill after all. Like porn flicks, police training exercises are not known for their in-depth plots or character development. The reaction from green groups was predictably shrill when the drill became public knowledge. Newspapers ran photographs of armed police crawling between cars with Greens stickers displayed prominently on their rear windows. In reality, the environmental movement prides itself on its non-violence, and arrests tend to border on the absurd; protesters cross an invisible line into the arms of police and then wait patiently on a bus to be processed. These arrests are counted by campaigners and announced to the media with much fanfare to show the lengths to which people will go to see a place protected or a policy changed. The process is usually so orderly that accusations of terrorism seem fantastical.

And yet the extremist tag sticks. During the campaign to protect the Franklin River in 1982, the state Liberal leader, Robin Gray, likened protesters to 'guerilla forces in third-world countries' while another minister demanded that all protesters coming from the mainland be stopped and checked for head lice. Rumours that the Greens were receiving funding from Eastern Bloc sources were rampant. Bob Brown recalls politely greeting a police officer on the banks of the Franklin River, only for the officer to turn his head slightly and say into his radio that 'Number 1' had been located. It would come as no surprise to today's campaigners if a deck of cards bearing their mugshots were circulating the island.

The 'eco-saboteurs' label, however, is not entirely unfounded. Its origins can be found in the writings of Edward Abbey. Notoriously contradictory and prickly, Abbey was a man of few spoken words who left a trail of wives and burning billboards across America. He wrote a kind of cowboy's poetry (when asked if he had ever come close to death, he answered, 'Got on a few horses I didn't understand'). His 1975 novel *The Monkey Wrench Gang*, about a group of misfits sabotaging machinery, spoke to the hearts of dispirited environmentalists. A year after its release, a group of burnt-out campaigners formed Earth First! Their catchcry was 'No More Compromise.' 'I'm all for compromise,' Dave Foreman, one of the founders, said at the time. 'It's just that our opportunity for compromise passed about one hundred years ago. We're down to the last 5 per cent of old-growth forest. We could have compromised at 50 per cent, but we didn't. We've got to save all that's left and begin to restore some.'

The group put together a handbook introducing activists to 'monkey wrenching.' Diagrams explained tactics such as tree-spiking, pouring sugar into hydraulics, cutting wires, diverting effluent back to its source, disabling animal traps and 'revising' billboards. The aim was to make polluting companies pay so dearly that their projects became financially unviable. The handbook quickly attracted the attention of the FBI, and Foreman was arrested for conspiracy. A decade later the FBI launched 'Operation Backfire' to tackle what it now called 'eco-terrorism.' It was seen by many as an over-reaction to what were mostly publicity stunts, such as rolling a giant plastic 'crack' down a dam wall, breaking into battery farms to film chickens drowning in their own shit, sabotaging traps and snares by leaving fluffy toy animals in their metal jaws, or pouring syrup into engines. But the group's actions, the authorities argued, held other people's rights in contempt.

Almost as soon as Earth First! began, mainstream environmentalist groups sought to distance themselves, worried that the radical group would undo all the inroads environmentalism had made into the lives of ordinary people. Yet radicalism has proved a siren song for some activists: there is always the temptation to say, fuck it, no more friendly logos, pretty pamphlets, strained meetings with politicians and industry; no more long nights deciphering bureaucracy with its endless loopholes. The Wilderness Society's Vica Bailey, a well-mannered, well-presented spokesman for his cause, admits the prospect of saying 'no more compromises' can be alluring. 'I'd love to go out to the forest and say "no more." Chuck out all the reports, the election campaigns and meetings.'

There is a tendency among some activists and onlookers to romanticise the forest blockades as the 'front line' and to dismiss more mainstream actions. One morning at the blockade, an older woman arrived with boxes of food. She placed the donations at Lunchbox's feet, like he was some sort of deity, and called out to the possums in the tree-sits, 'Thank you, Bravehearts!' Nathan shifted uncomfortably under her loving, uncritical gaze. 'Bravehearts?' I said incredulously, once she was out of sight. He grinned sheepishly and lugged the pumpkins to the kitchen.

Meanwhile, campaigners and scientists working in under-funded and under-staffed offices are accused of being sell-outs. Curiously, they seldom defend themselves, and instead praise the blockaders as heroes. Few environmentalists seem willing to go on the record to criticise the ferals; it's as if they've an invisible belt of protection around them. At rallies, Greens politicians call the blockaders 'brave' and 'heroic,' but in private, after a little prodding, Bob Brown admits that the feral element can be counter-productive. 'It's a campaign to save the forest, not a revolt against society,' he says. 'But what can we do?' He recalls that during the Franklin River campaign, they had to keep a constant eye on a particular protester with a green Mohawk who shot off outrageous statements at the cameras whenever he had the chance. Reining him in 'could have been a full-time job in itself,' Brown muses. In other parts of the state, more organised community groups have politely asked the itinerant protesters to leave their blockades. 'We felt horrible doing it,' says one woman from the Friends of the Blue Tiers community action group in north-east Tasmania. 'But they just bring a whole new element to a campaign that we didn't think would help us in the long run.'

Wazza gives the bluntest and least politically correct reason for why the 'feral' subculture is attracted to direct action. 'It's instant gratification,' he says. 'People say we're toeing the line, on the front line, putting our bodies on the line. All that heroic stuff, even I've said it. And to a degree, we are. But it's also instant gratification.' When I ask just how far the monkey-wrenching element goes, he shakes his head. 'It doesn't. It's more something symbolic to sustain us, something fun to hang onto.' That runs counter to recent accusations by Forestry Tasmania that the forests have been booby-trapped and that one logger almost lost an eye when protesters tied up the trees with cables, making it impossible to fell them safely. The timber worker said he saw a 'flash of metal' up in the canopy and stopped working just in time. But as no charges have been laid nor any statements issued by the police, it's difficult to tell if the accusations are just a ruse to smear the blockade. Any hint of Earth First!-style vigilantism gives industries and politicians a huge wedge to drive between greenies and the rest of society (not to mention a scapegoat for insurance claims and a convenient distraction from other political issues, especially during election campaigns).

The Black River bomb is one such example. On the morning of 11 March 1993, a length of wire leading to a homemade mixture of fertiliser and diesel was found beneath a railway track on the Black River Bridge in north-west Tasmania. A banner had been hung from the bridge reading 'Save the Tarkine: Earth First.' The absent exclamation mark was a glaring omission – that and the lack of a detonator. But it was two days out from a federal election, and the tabloids went wild. Polls had recently revealed that the Tasmanian Greens could win their first Senate seat and with it the balance of power. The bomb was a god-send for anti-green campaigners. Local newspaper the *Advocate* reported that 'eco-terrorist group, Earth First!' was behind the bomb, and timber-industry spokesmen and politicians ran with the story for the next two days. The Greens missed out on a Senate seat by less than 1 per cent of the vote.

Months later, Tasmanian police released a brief clearing the environmentalists. They also hinted that supporters of the timber industry were under suspicion of planting the bomb, their brief stating that 'innuendo in the Smithton community espoused the view that the incident was the work of the pro-logging community, the aim of which was to discredit the conservation movement's program during the summer months.' It added that there was no direct evidence to support this accusation. A Victoria Police memo was much more succinct, stating: 'The device is considered [by Tasmania Police] to be an elaborate hoax and they have not ruled out the possibility that it may have been placed there by loggers in an attempt to discredit the Green movement.' Loggers have since taken to plastering their trucks with bumper stickers reading 'Earth First. Log the other planets later.'

CONSEQUENCES

Sitting in the backyard of the Pink Palace, Christo Mills reluctantly agrees to tell me about the action he and the other blockaders put together a year and a half ago. He would prefer just to give me his legal testimony to read, but he hasn't been able to find it in the domestic debris. Nervously he rolls the same cigarette over and over.

It was a big action, one that had taken weeks to scout and plan for. Overnight the SWST crew built four obstacles for loggers to contend with before they could enter a coupe where thousand-year-old forest had been tagged for clearing. The first obstacle was Mills, lying in a hollowed-out car with his arm locked to a pipe cemented into the road. Tall and lanky, he practically had to fold himself to lie flat. He had a collection of rolled cigarettes, some water and a blanket, and five other protesters sat with him. Further down the road, out of sight, another obstacle was set up and then another and another.

At around 6 a.m., the logging trucks pulled up and less than an hour later, a couple of police cars arrived. Mills' support crew were promptly arrested.

'We know this cop, we call him Bipolar, cos he just swings from good cop to bad cop,' Mills says. 'But I felt okay because there was another cop with him, a female.

'Then he told her to go away. She came into my view and I pleaded with her to stay. But he sent her away.' From the back of a police car, the other protesters say, they could hear Mills yelling for her to stay. Some of the loggers started to jeer and laugh at him. The female officer got into the car, turned the ignition and they were gone.

A freelance news photographer, Matthew Newton, appeared on top of the hill around about this time.

'I got a tip-off there was an action, so I was the first media there,' Newton tells me. 'I came up over the hill and saw the car surrounded by a circle of loggers and the policeman leaning into the car. But when the cop saw me, he arced up and told me to nick off. So I headed back to my car and met up with the *Mercury* journalist on the way.'

Mills says the cop smashed the window over his head, unlocked the door and climbed into the car with him. 'It's normal during a lock-on to get a lot of crap, don't get me wrong,' he says. 'In a tree-sit a protester can just look over the edge and wave. But on the ground, you get stuff thrown at you, a couple times men have dropped jack-jumpers on us, stinging ants. People talk in your ear, spit on you, tell you you're scum. That's normal. But this was out of control. He was sticking his fingers in my eyes and up my nose, and slamming my head against the steering wheel. He put me in a headlock and kept saying, "I'm sick of you Greenie cunts." I could see outside of the car only a little bit, just got flashes of loggers leaning in and yelling stuff like "Look at the stinking hippy in the cage."' At one point, Mills says, the cop plucked Mills' wallet out of his pocket and passed around the photos of him and his girlfriend. 'Then someone reached in and pissed in my tobacco tin.'

As Mills is telling me this, I'm worried for him. His eyes focus on the floor. The cigarette in his hand has been rolled about seven times by now. 'They must have called up their wives or girlfriends to come down and have a look, cos suddenly a couple of them were there, taking photos of me. That was probably the worst bit. Some of the loggers started jacking the car up while my arm was attached to the ground. I was swearing because it hurt and they kept saying "Don't swear in front of the ladies," bashing my arm until I swore and then bashing me harder for it. Then I yelled as loud as I could in the cop's ear. It was a stupid thing to do. He pushed my face into the steering wheel and I think that's when my nose broke.'

Eventually the local fire brigade was called in to shear Mills loose from the car and he was taken to the police station. 'I sat up front and he played country and western through his loudspeakers on the roof. He kept talking to me, trying to befriend me. He said stuff like the Sea Shepherds were his heroes.'

At the station, Mills says, he was treated well by the other officers. 'They kept asking me if there was anything I wanted to complain about and took swabs from me, everything. But I just wanted to get away from all of them. I didn't know where the cop was, if their questions were a ploy to get him back in the room with me. So they processed me and then let me go.' In Hobart he went to a doctor, where his nose was re-set. 'We had no footage of what happened because he arrested everyone. Then the *Mercury* came out and said their journalist was there the whole time, so I was making it up.' Mills isn't sure why the newspaper came out with this statement but suspects it had something to do with them not getting the story first. But Matthew Newton says there was definitely a period of time when no media were present. He'd stopped to chat to the *Mercury* reporter and photographer on his way back to his car. 'I don't know how long we stood talking – maybe about fifteen minutes.'

After much urging from his friends and fellow blockaders, Mills lodged a report and an internal inquiry was made into the policeman he believes assaulted him. 'But when I saw his statement, I knew nothing was going to happen. Everything I had said, he had replied to. It was obvious he'd read my statement before he wrote his own. He said things like, "I had to use force at this moment because he was biting me." That kinda stuff.' Finally Mills lights his cigarette and looks up at me. He shrugs. 'My folks were down the other week. I really wanted them to meet him. I took them to the Florentine to see the forest but he wasn't there.' He smiles sheepishly before we both burst into laughter at the weirdness of it all. I imagine the many daydreams he's had playing that meeting out – his parents looking the policeman in the eye, trying to appeal to some humanity. It's a naïve daydream, and Mills seems to know it – what would a hardened cop care for a couple of middle-class parents and their ratty kid?

It is often said that the point of a protest is to demonstrate your willingness to accept the consequences. When I put this to Mills, he agrees. 'Definitely.' But even so, he looks rattled. Exhaustion scissors across his skin and there's a burnt-out-match tinge about him. When he isn't doing blockade 'stuff,' Mills works part-time as a carer for people with disabilities. In the yard of the Palace, he stretches out his legs, pushing off the dogs lounging on his heavy unlaced boots, and looks at me. 'It's strange, you know, us telling writers like you what happened. I don't even know if this is about what happened to me in the car or if this is just part of the campaign.' Inhaling deeply, he says quietly, 'It's just what happened.'

*

Most of the activists seem to forgive the loggers their violent lapses. 'It's not their fault,' says one blockader. 'They get psyched up by people telling them bullshit.'

'It's a shit feeling watching the logging trucks turn the corner, only to find us in their way,' says Bridget. 'All those clichés about putting food on the table and having mortgages, they're true.' She describes hearing the trucks first. The roar of the engine, chains rattling in the empty tray and fumes stuttering out of the exhaust, before headlights lit up like a pokie machine turn the corner towards the blockaders. The forests are sometimes so foggy they have to light fires along the road to stop the truck ploughing straight through them. Tripods poke eerily out of the mist, a body poised atop a cluster of enormous pick-up sticks. It is a spooky scene, the spindly structures and shadowy figures lit up by the headlights.

'The driver will most likely stay in his cabin and we'll stay on the road,' she says. 'Sometimes he'll come out and have a few words, but mostly no one says anything.'

By locking themselves onto machinery, the activists are putty in the hands of loggers who have been worked into a state of righteous violence, sometimes provoked by the state's own politicians. One protester had an angle grinder started up so close to him that the sparks burnt his neck. Even in a tree it can be dangerous. A female protester left on her own in the bush when her ground crew were chased off spent hours in a state of terror as two timber workers stood at the bottom of her tree, threatening to rape and hack her into pieces. Then they started setting up dynamite, yelling that they were going to blow up her tree. For some it has been hard to keep it together, while others shrug it off, saying, 'It's cool. I'm over it.'

In 2009, at the initiative of the Rudd government, a community cabinet provided a chance for Tasmanians to meet with ministers one on one or in small groups. Ula Majewski had a meeting with Peter Garrett. 'The last time we met,' Ula recalls, 'I'd handed him his balls on a platter of woodchips in a Hobart restaurant, where he was dining with other ministers. They were made of soybean protein and deep fried.' When they met again in Launceston, Garrett greeted her with a wry smile. 'Nice to see you again,' he said, while Ula searched for a glimpse of the 'old' Garrett. 'I showed him photos of some of the violence we've had at the blockade. I wanted him to know that things haven't changed out there. That things he would have experienced back in the day at protests still go on.' But she holds no great hopes about the meeting. 'Everyone just hand-balls you to someone else, and those that have the power to change the situation are too scared to do anything in case they lose that power. Sounds pretty useless to me.' Mostly she is eager to get back to camp. 'I worry about us,' she says of the SWST crew. 'A lot of really bad things happen out there and we don't have time to deal with it emotionally. It's almost at an endgame scenario for these forests right now. We just have to keep going.'

When SWST uploaded the footage of the attack on Miranda and Nish onto YouTube, it made headlines across Australia and overseas. Bob Brown compared it to the forest protests at Farmhouse Creek, where police gave angry loggers the nod to storm a blockade. An iconic photo of Brown was snapped as timber workers violently dragged him out, his shirt ripped and his arms splayed like Jesus Christ. But local commentators argued that today's activists were provocateurs, poking sticks at animals in cages; what did they expect when the loggers went wild? When I phone Rod Howells, the logging boss accused of leading the attack on the car with Miranda and Nish inside it, I listen as his wife hands the phone to him cautiously. 'A writer,' she hisses. Rod, however, is surprisingly less guarded. He tells me that the activists set him up, that the girl filming from the overhanging tree was waiting for him to lose it. 'I'd warned them. They knew they had a bit of a livewire. They set me up.' I know from photos that Rod is a barrel-shaped logger with a greasy blond helmet of hair, a small but stocky man. He says it was the third time in four months the activists had done an action against his logging business. 'I was sick of it. There was someone in a tree, on top of some sticks and now the bloody car dug in the road.' I ask him if he's watched the activists' video of the attack. 'Nup, I haven't seen it. My wife and my grand-kids have seen it, they sit around and have a bit of a laugh at it, but I haven't seen it. Don't need to see it. It's just a video of some windows being broken.'

Days after the incident, some SWST crew went back into the Howells' logging coupe to stop work. The group was split over whether they ought to go – after all, Howells' crew had made their feelings very clear. Nish decided to join the action. 'I didn't go,' Miranda says. 'I thought it would look bad if we did an action against them so soon after. Like we were asking for trouble.' Others thought it had to be done to show loggers they won't be scared off.

Controversially, when Premier David Bartlett was asked about the alleged violence against protesters in the forests, he said, 'I think the protesters need to take a good, hard look at themselves and make sure they're not impeding the legal work of forestry workers.' Local journalists had a field day with this quote, interpreting it as a green light to timber workers wanting to vent their frustrations on activists. But to a degree some protesters agreed. 'We know we are targeting the wrong people. It's Gunns and Forestry Tasmania who should be seeing us everyday,' says Miranda. 'But if we weren't here, all this –' she stretches out her arm at the green forest, the tinkering of birds emanating from within, 'would be gone.'

*

Later, sitting next to the road, Morton is adamant that time is running out for places like the Upper Florentine and that the block-ade cannot take its eyes off the forest for even a second. A largish Norwegian fellow with unruly red hair and dressed in a dark wool-len trench coat and waterproof pants, Morton stands up when a hired campervan appears on the road. 'I'll see if they want a cup of tea.' But as he puts his arm out, all I can think is that he looks like a goddamn bushranger. The van slows down, but the woman in the passenger seat quickly winds up her window. In slow motion they drive past us, staring at Morton, their mouths open, before quickly hitting the accelerator. Confused, my red-haired friend sits back down. Then in a flash of clarity he looks at me.

'I think we're focusing on the wrong thing, trying to show peo-ple the forest,' he says. 'We expect people to feel the same way we do about it. I think we should be focusing more on jobs.' When I ask him why he hasn't suggested changing the campaign's direction, he shrugs. 'There's no time. They want to smash this place bad. We reckon they know the tide is going to turn on them so they're try-ing to get to the best places before they can be protected.'

Morton says that sometimes, when he or the other activists are sitting here, a busload of Japanese tourists drives past the blockade. The bus slows down so they can take photos. Holding up their dig-ital cameras to the permanently sealed windows, these tourists must wonder why this motley collection of youths are in a forest in the middle of nowhere. Do they realise that the funny-looking kids are trying to stop this same forest from making its own journey to Japan in the form of woodchips?

*

One morning at the blockade I wake up early. In the quiet you can hear the cracking and clipping of abseiling equipment as the tree-sitters begin to stir. Padding softly into the open in my socks, shaking my boots out before putting them on, I walk beneath Miranda's tree. Instinctively I look for a telltale pile of scraps at the base, something to indicate her presence, as if she is an owl coughing up fur and bones. By the main road, the morning shift is on, blue smoke ribbons rising up from their fire. At the end of the unfinished logging road, past unearthed roots and drying lichen the colour and consistency of tobacco, the forest glows green.

Stepping inside it, it is difficult to be objective. A carpet of moss stretches over the floor and caterpillars creep along the branches. Man-ferns flirt around each other, trunks like brown furry necks seeking a patch of sky. In among the fishbone ferns, I watch as a tiny black bird with a hot-pink ruffled chest strikes a dozen sharp poses for his plain brown companion. Bright blue mushrooms huddle, and flames of orange fungi look like they've come off the side of a hotted-up V8. A red starfish mushroom stinks like rotten meat, making me retch. My favourite mushroom is the puffball, because it's interactive. A light tap on its side triggers a cloud of spores into the air. It makes me feel helpful.

Leeches sucking secretly on my legs, I try to see this place as a merchant might. As commercially viable timber, as furniture, as beams that do not bow, the skeleton of houses, as toilet paper, as floorboards, chopping boards, boats, fruit bowls; as paper, crayfish pots, pencils, souvenir barometers and placemats. Most of all I imagine the forest disappearing behind me and turning up in the office photocopier, or wrapped around a present, or in this book perhaps. Every now and then I come across the odd giant tree, its base like the foot of a dinosaur. These eucalypts splinter up into the sky so high that I have to lean backwards to take it all in. *It will grow back*, I am told by Forestry. Like hair, like fingernails, like skin. *It will grow back.*

LOGGERS

AT THE PUB

I'm at the National Park Hotel on the road to Maydena, the last town before the Florentine and Styx valleys. It is dark when I walk in and order a beer at the bar. Ignoring the hush, I try to act as if I always walk into pubs full of men in the middle of nowhere. Perched on a barstool, I flick nonchalantly through the paper – but as I scan the front page, I realise my timing is unfortunate.

'Isn't that the river I just drove over?' I asked the barman, holding up the front page. It shows a photo of an Asian girl alongside a picture of the Tyenna River. The barman jerks his head.

'Yes, but it wasn't any of us.'

A voice behind me chips in. 'Definitely not us. That's a stupid place to put a body. I mean, there's hundreds of mine shafts around here.'

I turn around, but there are too many eyes looking at me expectantly and I can't figure out who spoke. I nod slowly and read through the article. Police have arrested two Hobart men after discovering the girl early this morning, her body weighed down by rocks in the shallows.

'Are you a greenie?' another voice calls out, startling me. I spin around quickly this time and catch the speaker, a young man wearing an orange hi-visibility top and blue pants; he falls back behind his mates as if to share around responsibility for his question.

A couple of girls have walked in to buy a six-pack of sweetened mixers and they pause to look at me. I shrug.

'I dunno. Are you?'

His mates semi-shriek and fall over themselves, while he puffs himself up.

'No way!'

I tell them I'm a writer and that I've been staying at the Florentine blockade up the road. The men recoil.

'They stink, don't you reckon?' says a fella in a hi-vis fleece jumper. 'They smell disgusting.' I don't answer. 'Go on, admit it, they stink. If one were sitting right there, would you sit next to them?' I say I would and the men scream with horror.

'No way! Go on, admit it. They smell,' another pitches in.

'Well, yeah,' I begin tentatively, 'they've got their own …' But before I can finish, they're clapping and cheering.

'She admits it! She thinks they stink!'

I start to laugh, giving in. 'Okay, some can smell a bit gross, but I reckon Lynx deodorant and most women's perfume stinks as well.'

Some of the men nod sagely. Others are giddy with joy that they've made me say it.

I look around. It's an okay pub. Stickers are plastered all over the walls, the usual 'I love beer, shooting and fucking' jokes ('Credit is like sex – some get it, some don't,' I spy above the door). It's perhaps a little too antiquated, cashing in on the 'Oh, aren't we oddballs, us Tasmanians' theme for the tourists. But at least it makes an effort to charm, unlike the other pubs I've seen on the island, renovated with rows of pokie machines and shrieking metal chairs. I look up at the dinner menu on a blackboard. I want to order a vegetable pattie but worry it will condemn me to 'greenie-yuppie' status. Asking for a salad roll can be fraught with danger down here.

When I asked for one at a café in New Norfolk, the woman behind the counter looked at me with suspicion and asked, 'You want ham, chicken or beef with that?' I order fish and chips, silently apologising to my stomach. I hate meals that are all yellow.

When my meal arrives, a parmigiana is plonked down in front of the man beside me. He introduces himself as John, and tells me he is a tree-faller, a subcontractor in a harvesting team in the Styx Valley. Fallers are among the few who still touch the trees before they fall. If a tree is too difficult to cut with the jerky outstretched arm of a machine, the faller steps in with a handheld chainsaw and with a series of cuts can make it fall away from the crew. 'You can tell the fallers,' one activist told me earlier. 'They're the skinny ones. The rest are fat from sitting in machines all day.' John isn't skinny but he isn't fat either.

'We're also the "sawlog chasers,"' he explains; they trim the tops, branches and butts off felled trees and separate any sawlogs from the stem – the straight, wide logs suitable for boards rather than pulping.

A big man with blond hair, John is in his early thirties. He was a shearer before he got into falling, but his current occupation runs in the family. 'My grandpa did it, my father too, but he got out after Gunns took control of everything. Said it was too much of a monopoly.' John stayed in the game. His family lives in Buckland on the east coast, a few hours' drive across the island from the Styx Valley. He commutes to work, living at this pub during the week and at home on weekends. 'I'm trying to get the boss to give me a portion of his fuel subsidy,' he tells me. 'But it's not looking good.' John is a thoughtful presence amid the fluoro rowdiness. As we chat, there are times I think he hasn't heard me he is quiet for so long. I ask him about the activists.

'It's the weirdest thing,' he says slowly, two beers between us. 'You come to work and there they are. One dressed as an eagle, another looking like she wants to tear your eyes out, and then they try to talk to us as if this is normal.' John is part of a small 'bush crew,' contracted by Forestry Tasmania to fell a coupe and haul the timber to a mill. '[The protesters] tell you to call the police and FT, so you have to drive back in to phone reception, and then the boss tells us to go back – and then they want to talk to you all mate-like. Offering us cups of tea and bullshit like that. I mean, fuck, we're here to work, not to have a chat.' He shakes his head and says again, 'It's the weirdest thing. You see them barefoot on top of a machine or tree, and there's a frost and they're dancing round, singing – they're fucking crazy.'

For many of the activists at the Florentine blockade, it's wood-chips they have the biggest problem with, the turning of whole hectares of forest into mulch. I am told repeatedly that sawlogs account for only a small portion of the industry's output. They are cynical about the industry's claims that sawlogs are their top priority.

When I put this to John, he looks genuinely confused.

'It's the cream of the crop. I'd have to be stupid to send a sawlog to the chipper. What's left over goes to woodchips.'

I tell John what the activists at the Florentine told me: 80 to 90 per cent of the forest felled there will go to the chipper.

'Yeah, but sawlogs are the prime cut,' he insists.

Like a carcass in an abattoir, the forest is divvied into cuts. But unlike a cow that couldn't live without its T-bone, perhaps a forest could be left standing and just its prime cuts removed? John considers this for a while before answering cautiously.

'In regrowth forest, selective logging would be possible, but in old-growth it would be way too dangerous. You'd have to clear-fell.'

John already has the most dangerous job in the industry, and the most under-represented. When I ask him who in all the forestry brouhaha speaks for him, he again goes quiet, as if fishing in some dark place for an answer.

'Not Ferdie Kroon [CEO of the Forest Contractors Association], that bastard loves himself. And Barry Chipman, he's a wanker. I mean, who is Chipman? He is quoted in every bloody news item but I've never seen him [Chipman is the Tasmanian spokesman for Timber Communities Australia, an industry group].' John laughs out loud. 'And now Brant Webb is running for election. What a goose.' Webb became famous when he and another miner were trapped in the Beaconsfield mine for two weeks. Their survival story made headlines around the world; they even made an appearance on *Oprah*.

As we talk, advertisements play on the television above the bar. There's one promoting Forestry Tasmania, another publicising a 'Walk Against Warming' to be held in the Upper Florentine. A tourist on a nearby barstool overhears our conversation and starts to talk to the man behind the bar about the logging of old-growth forests.

'We used to go into the pub and say we were loggers,' John tells me. 'Now we keep quiet if we don't know the place. It's intimidating. I wanted to go see Johnny Diesel play in Hobart last month but he was at the Republic Bar and I just thought I'd stand out like a sore thumb.'

Around us, there's a hum in the room, fuelled by talk of timber. A younger logger leans between us to listen. Boyish, he gains confidence as more men gather. Pulling out a mobile phone, he shows me videos of working in the forest – skidders going back and forth, timber being dragged to a landing, trees falling over.

John eyes him and says with a dry smile, 'When have you got time to do this, Mitch? Shouldn't you be working?'

Mitch laughs and shows another clip to a man standing near us.

'Oh, I reckon she would have already seen that one, mate,' the man says knowingly. Curious, I lean over. It's the video of the car being smashed with Nish and Miranda inside it. Mitch must have downloaded it onto his phone. I wonder how many times he's watched it. He giggles as replays it for me. I look at the faces around me, wondering who might be one of the blurry bodies in the video. I ask John what he does when protesters come into a coupe he's working in.

'I try to keep out of it. Just don't say anything. But there was this one time we'd started logging a coupe in the Styx, 10F, and this woman she came out of nowhere and was going crazy at us. She was sobbing, a thirty-something-year-old woman, sobbing over a few trees. I was pretty rude to her, I could have dealt with that better.' Again I ask John about clear-felling.

'It's the safest way in the old forests,' he says. But what about all the potential sawlogs that go to the chipper before they've grown, I say. Isn't that your future yield? John is quiet. But then Leon, the pub owner, appears behind the bar. Leon is a retired logger himself, and John snaps up his head.

'It would be okay if you guys didn't cut everything in the '70s.'

A small round man with white hair and glasses, Leon barks back: 'That's not true – we did a good job around Triabunna. All the wood was rotten, so we cleared it and reseeded.'

John shakes his head accusingly. 'That's bullshit. There was good wood in there.'

'No, it was all rotten, John, I'm telling you.'

'Good wood and you guys cut it all for the chipper. So now we're two generations behind.'

I'm not sure what I've started. The man sitting on the other side of me winks and says in my ear, 'Who needs an activist?'

Further down the bar, another man puts his beer down and says loudly, 'I recommend that everyone shut their mouths and not say another word to this woman.'

There's a silence. I lean forward to look at him properly. He's wearing a fluorescent orange jacket, and has closely cropped hair and a pointy face that almost fits into his beer glass.

'She'll take it all back to the Greens, to Bob Brown. Or you'll be on *A Current Affair* next week.'

I protest. 'C'mon, mate – I've got standards. *Today Tonight*, at least.'

The man beside me says, 'Lay off, Turk, she's okay.' But the others are quietly looking me over more closely.

'So, whose side are you on?' asks John, leaning back, his blue eyes on me.

MAYDENA

Maydena's main street is empty and most of the curtains are drawn. There is one fence that is out of the ordinary. Anti-forestry slogans are painted on it and two Ford panel vans sit in the yard, covered in hand-painted pictures of Aboriginal flags and native animals. A whiff of aniseed lingers, and a vegetable patch has been ploughed in rows next to the footpath. I have come to talk to Prue Barratt, one of the original Florentine blockaders.

'Someone tore down my first election poster of [Greens senator] Christine Milne, so I ordered three more and hammered them on the roof,' she tells me. Prue recalls her mother having similar difficulties. 'Once she put a Greens poster in our front yard and it disappeared. Then months later at a neighbour's barbeque she found it pinned up in the garage, covered in dart holes.'

Prue obsessively thinks up new ways to foil her saboteurs. After star-picketing the signs in her front yard, she covers their edges with ink so it will show up on people's hands. She has even constructed a fake electric fence around one particularly abused poster.

'But ten-year-olds still stand in my driveway, throwing rocks on the roof, yelling out, "Log the fucking Styx" in their squeaky voices.' Every three months or so, her fence is pulled down. 'There's been the odd drunken idiot coming down the drive waving a stick or something, but most of the men around here are just confused by me.' Prue laughs. 'They don't know whether to shoot me or fuck me.'

A kind of bogan embodiment of Eastern philosophy, Prue swears prolifically, sells organic vegetables out the front of her house, is a strict vegan, chemical-free ('apart from toothpaste') and determined to live alone. Petite and handsome, she pumps iron in the afternoon, working on her upper-body strength.

'They were supposed to give in years ago,' Prue says, her tiny hands slapping her thighs. She was born with her feet turned backwards. 'My mum had to carry a wrench just to change my nappy,' she jokes. Her father massaged her feet every day until in time her parents had manipulated her feet to the front. After a stint in built-up shoes, Prue could walk without drawing attention to her condition. These days, easing off on her strict consumption rules, she sips home-brewed stout in the evenings for the pain.

'I'm fourth-generation Tasmanian,' she tells me proudly early on. Both her grandfather and father worked for the local paper mill, Australian Newsprint Mills, an operation largely instigated by Keith Murdoch. ANM built the town of Maydena back in 1947.

'My grandfather surveyed the Florentine for ANM and divided it into the coupes that we're now trying to protect,' Prue muses. At forest actions she wears her father's old ANM jacket. 'He doesn't mind,' she says. The only family member who does mind is her brother, a forest contractor. Although he lives close by, they haven't seen each other for years. 'There are lots of reasons for that,' Prue explains, 'me being a "stinking fucking greenie" is one of them, but not the only reason.' She concedes her being on the 'other side' made his work difficult. 'His co-workers said things like, "She's *your* sister – how's she finding out about these coupes or where the 1080 poison has been laid?" Shit like that.'

Past election results in the town of 200 people reveal that Prue is not as alone as she thinks. In the state election before last, thirty people in Maydena voted for the Greens. But few are coming forward to say so. Later I meet a local woman who admits to me she is one of the thirty. 'But I am a mother,' she explains. 'I've my family to think of, my son has a hard enough time here already. I'll vote that way, but I can't say so.' Down the road, John MacCallum, a hostel owner, says, 'Being a greenie here is like coming out of the closet. Worse even.' Home to a mixture of misfits, loners, war veterans, retirees, low-income and welfare-dependent families, plus a few small-business owners, truck drivers and the odd timber contractor or shooter, Maydena is barely recognisable from its glory days. A policewoman was recently installed in a small outpost off the main street; inside, her three cats lounge on the counter. Before her arrival, vigilantism was common.

'We burn them out,' one punter at the local pub tells me. He talks about junkies who move here for the cheap rent and the handy lack of job opportunities. 'Perfect place to lodge a dole form – there's no work round here.' He says there's usually a spate of robberies when one of these families moves in. 'So if they steal from us, we smash down their door in the middle of the night, give them a fright and by morning they're gone.' When I ask if they ever get it wrong, he shakes his head at me. 'There's barely 200 of us, so when a junkie family moves in and one of us gets robbed, it's pretty obvious who's done it.' Later, another resident, Ali, a friendly young woman with freckles who runs a guest house about half a kilometre down from the pub, confides, 'Look around you. Everyone's got a big black dog in their yard. And a white ute. So we got one of each too.'

Prue is the only local who tells the town what she thinks – in monthly instalments, in house paint on her front fence. 'I don't know whether to think she's brave or crazy,' Anne Nevin says admiringly of her fierce neighbour. Pointing out a platypus paddling across their pond, its sleek coat moving like a tiny oil slick through the water, Anne also manages accommodation for tourists and visitors. She explains that her family cannot afford to pick sides. Both forestry and environmentalist groups rent her guest cottages for conferences. 'Maydena is such a beautiful place,' she says, 'but it has been hard for us.' From rural England, the Nevins were drawn to the area because it reminded them of home. The similarities, however, ended with the velvet green hills, imported weeping willows, cold sunlit mornings and footprints in the frosted grass. When Anne had a letter published in the *Mercury* asking forestry and environment groups to attempt to mend their thirty-year rift, she received phone calls with nothing but the sound of a gun going off at the other end of the line. 'A few times people would book dinner here for ten people and never show up.' Her daughter has found shotguns under mattresses when changing guests' bed-sheets.

Behind the Nevins' workshed, the 'Welcome to Maydena' sign is lying on its side, gathering cobwebs and rust. In telltale fluorescent paint (the same type loggers use to mark sawlogs from woodchips) someone has written across it 'Fuck Off Green Scum.' 'My husband got a ladder and removed it from the main road as soon as we saw it. No one has offered to help clean it and put it back,' she tells me.

In a town where you could know everyone's name, Maydena seems a lonely place. Curtains are mostly kept drawn, the blue light of televisions shining under their hems. When I visit the community centre in the primary school, I find one resident sitting at a computer, looking at clothes on eBay. Locals seem used to broken promises. When Forestry Tasmania announced plans for a 'Maydena Adventure Hub' five years ago, John MacCallum recalls,

'It was like the town had come to life. I thought they had all given up, rotting in their houses.' But it has been a long time coming. When FT finally launched the enterprise late last year, special guests and journalists were driven to one of its main attractions, Eagle's Lookout, where they looked out into a thick layer of white fog. 'And when it's clear,' quips a local who asks not to be named, 'they'll just see a carved-up landscape with burnt holes everywhere.' Reluctantly, MacCallum admits he has doubts about the venture and its never-ending stream of shiny brochures promoting sustainable forestry in the area. 'I think they will cut down the last tree. And I mean tree in the true sense of the word – not these straight, genetically designed trees that don't have arms ... but the last true tree.' He takes me around the front of his house and points at the bare paddocks in the distance. 'Look over there. They razed the forest and left two blackwoods. I don't even know what that means.'

Maydena wasn't built so much as unpacked – a prefabricated town. Today many homes still share identical fittings, door handles, windowsills and latches, front doors; its flat-pack design gives it a temporary feel. When ANM was booming, the town was home to around 3000 people, mostly employees and their families. The school was full and a town hall was built for dances and plays. Flush with cheap hydro-electricity and seemingly endless forest (the state government annexed 2000 acres of Mount Field National Park for the newsprint mill), the town seemed well placed. When ANM shut down in the early 1990s, the company returned its permit to log the Styx and Florentine forests. Some observers hoped the land would now be returned to protected status. Instead, the area was placed in Forestry Tasmania's care, allowing them to on-sell the trees to timber companies. Despite the millions of dollars now passing through Maydena at all hours on the back of logging trucks, these royalties barely touch the town. And yet locals seem as loyal as ever to the midnight haulers.

'We probably spend the most money here, buying hot chips,' Bridget joked nervously as we drove out of Maydena. Her little yellow car had no doubt been clocked a dozen times by residents peering through their windows, but still she drove furtively, foot on the accelerator as soon as the speed-limit sign reared up on the road. There is only one shop in town, the last place to get petrol and food for miles, but Bridget didn't stop. It is poised on the edge of Maydena, an al-foil gateway to the Styx and Florentine valleys. Out the front, the owner – a rotund man with thick lizard skin – stood on a cement platform and oversaw the road, unused petrol bowsers rusting on the concrete behind him. In the following months, he became a permanent marker for me. Like a spectre, he watched us drive through. His blue eyes locked on me and I couldn't help swivelling in my seat, neck craning, unable to break his gaze.

*

'So, whose side are you on?' John repeats his question as I flounder under the gaze of the men in the National Park pub.

'Ah … no one's,' I stumble. John releases a tiny sigh.

'Well, what do you think then? Should we keep logging in native forest?' Like spectators at a tennis match, the men follow our words back and forth over the bar. 'Um, I'm not sure. Definitely not in some forests.'

A couple of the men grizzle.

'You must have an opinion, otherwise why would you be here?' John presses. I shrug and nod at the same time, an involuntary and contradictory gesture. I give in a little. 'Okay. I like nature. I like creatures. I think they deserve more rights than they have right now.

I also like timber and I'm a writer. My whole career is built on paper. So I can't *just* have an opinion.' As I say this, it finally strikes me what's been pissing me off about some of the ratbags I met out at the blockade. Why aren't they in here having this conversation? Don't they need more than just a gut instinct to save the trees?

When I asked a local journalist in Hobart what he thought about the activists at Camp Florentine, he sighed. 'They alienate people. Educated people try to have an open mind about it. Despite decades of having our minds clouded by forestry, we try and keep an open mind and we want to understand it. The ferals don't help explain the case. People say, "Why are you up there? What would you be doing if you weren't here?" Whereas you look at people like Geoff Law, Peter Cundall, Richard Flanagan, Christine Milne, you know what they'd be doing if they weren't there. They have other things to do.'

Echoing this, John puts his beer down and says, 'You know what will happen if they save the Florentine? They'll pack up their shanties and go protest somewhere else.' It's the rootlessness of the ferals that people don't seem to trust; their claims of connectedness to all wild places touches a nerve. Even residents of Maydena who want to see the Florentine protected dislike the ratbags' itinerancy. 'It just doesn't feel genuine. More like an excuse,' one local woman says. The journalist had told me, 'They don't help. They muddy the whole view. I can understand the zeal but they don't help me understand the issues. I have to be convinced of the arguments in a more learned way than that.' And now at a pub in the middle of nowhere, I finally put my finger on some nagging feelings about some of the ferals and about myself. I look at John.

'My gut tells me that something's not right with forestry here, but I can't work on gut alone.'

Surprisingly, John changes tack and soon we're talking about the bigger issues.

'It's like I'm two people. Yeah, on the one hand I'm a logger, but on the other my aim is to build and live completely off the grid. You see, I don't like pigs being kept in those small cement cells any more than you probably do, but what can you do? You can't cut off someone's income.'

'But what if that man is cutting off another man's income because he wants to keep his costs as low as possible? Or using steroids or hormones to grow his pigs faster so he can undercut other pig farmers?' I ask. Some of the men start to wander away from us. We're beginning to sound like a philosophy tutorial.

John doesn't answer. He looks at his plate and is quiet. After a moment, he looks up at me and says, with a degree of finality, 'We do a good job out there.'

And suddenly I'm sad. Perhaps the man who told everyone not to speak to me was right. Part of me wishes I'd left John alone. I probably will betray these men, at least that's how they'll see it. They deal in black and white, in absolutes and taking sides, while I'll write a flimsy watercolour.

Peering over my shoulder at my notebook, one of the loggers says wistfully, 'There are times when I wish I could put pen to paper.' He says he'd like to defend his job against greenie attacks. 'But I wasn't the brightest in school. My teacher told me in year nine that I may as well give up the whole learning business and go out to work with my dad.' No one is very bright at fourteen, I say; maybe he shot through too quick to tell. But he is adamant that school held nothing for him and that he was glad to go. 'I was rapt. I wanted out.'

Eventually the drinkers start to leave for the night, walking home along the defunct railway sleepers. Leon starts wiping down

the tables and I walk out with the men as they have their last cigarettes for the evening. In the car park we shake hands. John says
he's coming to Melbourne soon to see AC/DC.

'Get out!' I exclaim, pushing him into Mitch. 'Tickets for that
sold out in less than two minutes.'

He nods, grinning. 'Sure did.'

We make arrangements for him to call me for a beer if he and
his lady have time, and then they stub out their cigarette butts and
we say goodnight. They go back into the pub, where Leon has
already turned out the lights. I walk back along the road to where
I'm staying, black dogs on a black night barking at me.

*

'Well, you would've been fine, you're not a feral,' Ali says to me
when I tell her I got on fine at the pub. She's has a pommie accent,
a friendly laugh and sharp milky teeth. 'They would have been on
their best behaviour. Then again they know me, know I'm not a
feral or part of that activist mob, but that didn't stop them smashing bottles on my car and threatening me. And I've a white ute,
black dog and everything!' She laughs. I've stayed the night in one
of Ali's guestrooms. She flounces across her kitchen, cats meowing
out of her way as she finds a teabag for me. Plonking the bag in hot
water, I ask her about the cats. There are a few of them. My eye is
on a patchy, skittish-looking kitten that doesn't want to be held. As
I lunge for it, it jumps from the couch to the kitchen bench and
into the sink. Ali laughs and tells me about a black kitten across
the road. 'The litter was all white except one, a little black runt.
Anyway, the mum was bit by a snake and all the kittens died from
sucking on her teats, except the black one because they wouldn't
let her suck.' Her blue-grey eyes shine. 'Isn't that amazing?'

I find it strange that Ali lives out here. She seems so open and warm next to her wary neighbours. It seems lonely.

'We have a connection here,' she says. 'My husband's grandfather, he was sent here to work at Butler's Gorge. He had come over as a choirboy in the Vienna Choir, sang at the Opera House and then Vienna joined the Germans so the entire choir and their conductor were kept here as a POWs.'

I'm aghast. I didn't know we had prisoners of war.

'Oh yeah,' she says. 'I think you had more than a few.'

I shake my head, jotting it down.

'After the war, when they were freed, there was no point for most of the boys to return home to Austria – there was no one left. Instead he came to work here, in the south-west of Tasmania. So it can get a bit frustrating when people tell us to butt out of the island's affairs, that we've no right to an opinion, when my husband's own grandfather worked his butt off for this place.'

There is so much memory on this island, I think. It seems to sustain people, even when communities don't.

I ask Ali what she meant about bottles being smashed on her car. 'Well, I'd called the police, but they said they were too busy at the blockade and the girls should have known better than to go to a loggers' pub.' During a rally, two female activists got the idea of going to the National Park pub for a six-pack of beer. When their car failed in the car park, some men came outside and started to abuse them. 'They started smashing bottles on the car, pulling it apart. One of the girls got punched in the face, called things like "You stupid stinking feral cunt, disgusting bitches." Anyway, so the girls run up to our place for help and I called the police for them. There were at least thirty at the rally, twenty minutes down the road.'

When the police ignored Ali's pleas to get the girls and their car out of the car park, she drove down herself to winch them out. 'I mean, whatever the police's opinion, or mine even for that matter, I couldn't just leave the girls in that situation.' By the time she had the girls' car out of the car park, several bottles had been smashed on her own bonnet. Men leaned in her window, telling her to mind her own business.

The police later phoned her back to apologise. 'Then they asked if the girls could stay with me. I was like, we're going to get burnt out, why won't you help us?' A few days later, Ali demanded there be a police investigation. 'It turned out the two girls had nicked the car from another really lovely activist. I mean you can't win with either of these people!'

She laughs – exasperated, but still laughing. Ali has a teenage son with autism and a young daughter.

When I visit her again a few months later, she's tired. There are streaks of grey in her hair.

'I was a mess for a while there, but I'm on the up now,' she says determinedly. 'I was driving and thinking that they can't wreck the one good thing about me. I'm friendly, I'm not going to stop just because everyone else is angry.' She tells me things have been tough. Her son has been getting older and stronger, trickier to handle, and business has been slow. 'Leon from the pub called me and said he had an overflow of bookings so could I take some of the forest workers in – which I could. So these fellas, they came down, picked their beds and dumped their stuff, took a piss and went back to the pub for dinner.' About to go to bed, she got a call from the pub. 'Leon told me our rooms weren't good enough and the men are going to stay at the pub. I was taken off-guard and responded on autopilot I think, I was polite and said of course, that's fine. But when I put the phone down, I burst into tears.' She went out to check the rooms she'd cleaned earlier in the day. 'I just wanted to look at the rooms and see why the men thought they weren't good enough.'

She bumped into some of the workers as they were getting their bags. They didn't say a word as they left. When she saw the rooms, Ali started crying again. 'The rooms were trashed. Not on purpose – but the men, they were in their work gear, there was mud and ash all over the floor, they'd pissed on the toilet seat and even where they sat on the beds, they left an imprint of dirt.' Ali says it took her a few days to realise that them leaving probably had nothing to do with the state of her rooms. 'What probably happened is that [Leon] had a couple no-shows and instead of saying that's life, he took the guys back.' Ali smiles bravely. I feel horrible for her. I think about all the men in that pub, trying to imagine why they wouldn't think for themselves, why they wouldn't just say, 'Shuddup, Leon, I don't have to sleep in the same place as my mates.'

The Turk was right. I'll betray these men and the ratbags too, not because I'm choosing sides but because loyalties run too deep here. No one will give an inch, let alone stand up for a woman with two kids and a smile for everyone.

On bumper stickers I keep reading 'Tasmania – Your Corrupt State,' but words such as 'corrupt' don't seem quite right. If anything there seems to be an indignant kind of mateship here, a loyalty that precludes empathy, and a pragmatic approach to the rules: 'Oh, screw all this red tape, he's my mate, he can do the job, I'll phone him up right now' way of going about state affairs.

*

As I drive away from the Florentine blockade and Maydena, over the Tyenna River, where police are looking for clues to the young girl's death, my heart slowly sinks. I feel like I'm stuck at the bottom of this murky debate and I have to find a way back to the surface. I think about Crazy John's accusation on the ferry over: am I a big-picture person or a writer getting stuck on the small things? Is it possible to separate the two on an island as small as this?

When Australia's federal minister for climate change, Penny Wong, was asked why her department puts pressure on countries in the Asia-Pacific to stop clearing their native forests while Australia continues to log its own, she replied tersely, 'We have commitments.' And she's right. Everyone seems to have commitments. Tree-faller John is committed to monthly mortgage repayments on his family home, local business owners have beer to sell and guest beds to fill, and timber contractors have quotas to meet. The blockaders have shed themselves of most other responsibilities in order to be totally committed to the forest.

Economic commitments provide security for investors and reassure voters, but Tasmania's constant committal of its ancient forests to the chippers seems increasingly absurd and at times deeply irrational. Pro-forestry decisions have flown in the face of mounting public opposition, concerned economic analysis and significant scientific evidence. But to let the island's forests stay standing, say timber spokesmen, would lead to a different kind of erosion, an economic one. The island's scaffolding, they say, is its timber industry. I wish I could find a remote control, a pause button, to see what would happen to Tasmania if logging stopped, if the ground stopped shaking just for a moment. Would the island buckle and sink, like the doomsayers predict? And considering Tasmania is the second-most welfare-dependent state or territory in Australia and has the second-lowest rate of school retention after the Northern Territory, just how sturdy is the island's economy anyway?

'We're not stupid,' said Jess Wright, a former Florentine blockader. 'We know we won't save the forest by hanging banners from branches. What we're trying to do is press pause so this place has a chance to survive while the people with power can think about what is happening here.' But there is no remote, only a foot wedged in the door, a flimsy blockade on the frontline.

THE SAWMILL

Tony Jaeger is drawing me a picture. I'm in his office at the McKay sawmill in Glenorchy, a sprawling suburb behind Hobart. Established in 1950, McKay runs three production sites across the state. Jaeger is general manager. The open-plan office is divided by waist-high varnished yellow wood partitions and banisters; it looks wholesome, straight out of a timber yard in 1970s North Dakota.

Jaeger grew up in the timber industry – his dad taught him how to cut a log as soon as he was able. 'My father used to take me to work with him at 4 a.m. because it was easier for me to get to school from there,' he says. 'He'd give me a piece of chalk and I'd walk around the site drawing on the trucks, the trees and the saws. Sawing is second nature to me. You have to be able to look at these trees and know how to cut them just from looking at them.' At a wooden table covered with clear protective plastic, he is showing me how trees have to be cut in the southern hemisphere.

'No one has been able to work out why, but in Tasmania, we can only cut our logs on the quarter. People have been trying to crack the code for years, but I don't think there is a code. I think it's the latitude.' The Tasmania quirk of cutting on the quarter, he says, means that there will always be markedly fewer sawlogs than their by-product, woodchips. 'To top it off, the boards shrink when we dry them. So while we may have recovered 44 per cent of the log, it comes out of production and is measured at three-quarters of that original size. So on paper it reads that we have only recovered 34 per cent of a sawlog while the rest goes mostly to woodchips. When the Greens say, "80 per cent of old-growth is going to woodchips," they don't understand that without maximising on the woodchip by-product, the costs just to get the sawlog out would be too high.'

Months later, I learn that recent improvements in the timber-drying process have made it possible for a portion of the island's eucalypts to be back-sawn – a method many loggers prefer for its convenience and greater yield. But much of the timber from regrowth forests is still prone to degradation if it is not cut on the quarter; back-sawing remains not only inconsistently useful but also, according to Jaeger, economically unviable.

'So, do the woodchip companies come here to get their chips?' I ask. Jaeger smiles a little.

'No, they get their chips from the forests or plantations. But what I'm saying is that we can't afford to go in and get the sawlog without getting the woodchips.' So, the argument seems to go, it's the by-product that makes the product viable. In the beginning, wood-chipping was promoted as an act of frugality, a way to make use of the 'rubbish' left on the forest floor after loggers had been through for the sawlogs. It wasn't until the 1960s and '70s that Australian timber companies began to focus on woodchips and clear-felling. Great big shaved patches of land appeared as the new industry took off. By 1972 four national companies had contracts worth about $460 million for 30 million tonnes of woodchips.

The role of the government has been crucial to the growth of the woodchip sector: prime ministers Paul Keating and John Howard both responded to pressure from unions and industry by abandoning limits on the volume of native forest woodchips that could be exported. In 2007, according to the *Wood Resource Quarterly*, Australia exported a record quantity of woodchips, over six million tonnes, about 70 per cent of which was eucalyptus, most from native forests. Tasmania manages to export more native forest woodchips than all the other states and territories put together, despite making up less than 1 per cent of Australia's landmass.

At a Senate hearing in June 2010, Paul Morris, the deputy executive director of the Australian Bureau of Agricultural and Resource Economics (ABARE), said that Tasmania's native forest industry is heavily dependent on the export woodchip trade. 'About 71 per cent of logs harvested in Tasmania actually go to the export woodchip trade. If you look at native forest harvest, about 80 per cent of the native forest timber goes to the woodchip trade, and about 93 per cent of the plantation hardwoods are going to the woodchip trade.' Forestry Tasmania's own data from 2004 shows that around 85 per cent of the state's forest harvest goes to woodchips. Forestry Tasmania operates as a 'government–business enterprise' and manages the state forests for logging and a few tourism ventures – but woodchips are overwhelmingly FT's main earner. Today, there are four woodchip mills in Tasmania, all owned by Gunns, including one at Bell Bay (north of Launceston), also the site of the company's proposed new pulp mill. At the Triabunna wharf on the east coast, enormous piles of woodchips undulate like sand dunes, dwarfing the workers who climb them. At Burnie, on the north-west coast, residents refer to the mounds of woodchip as their 'golden pyramids.' Locals read these piles of woodchip for news; their changing height is like a heart monitor for business. When the market is good, the chipping machines can process up to 1000 tonnes per hour, twenty-four hours a day, seven days a week.

When I was in primary school, the playground bitumen was replaced with woodchips. It immediately became our weapon of choice; the sick-bay teacher went from picking gravel from our torn-up knees to urging us to blink to get the wood particles out of our eyes. We never gave much thought to where the wood came from; it looked like it must be the remainder of something, like leftovers from some other job. But mostly we just enjoyed the loose earthiness of it, the seemingly never-ending supply, and happily threw it around.

Woodchips are also an accountant's dream: cheap public forests, a processing method requiring little delicacy or skill, quick turn-around and a seemingly insatiable market – Japan. Between 2001 and 2004 Tasmania exported around 4.5 million tonnes of wood-chips to Asia each year, mostly to Japan's paper manufacturers. While these volumes have since declined to approximately 3.4 million tonnes annually, that is still a lot of *Hello Kitty* stationery. Many activists consider it ironic that Japan, a largely forested island, conserves its own trees while importing native woodchips from the least forested country in the developed world. Accusations that Japanese companies 'rape' our forests are unhelpful, however. It could be argued that Japan has fuelled Australia's economic success; the small country is more likely to see itself as a benefactor than an exploiter of our resources. The timber workers I meet, on the other hand, show even less respect for their product's purchasers than do the activists. 'Fuck the Japs,' I'm told repeatedly by timber workers and industry spokesmen. '*The Sea Shepherd* are my heroes. I'm right behind those guys. But leave us alone – we're sustainable. Fuck the Japs.'

How did a product that began as a thrifty way to use up waste products become so central to the timber industry? More often than not, felled trees have defects in their branches, or the pith – the cork-like heart of the tree – is rotten down the middle. 'These end up going to the chipper,' Tony Jaeger explains. Government legislation guarantees the timber industry a minimum of 300,000 cubic metres of sawlogs each year, 'But we need to access more than that to get that amount of appearance-grade timber ... We need bigger-sized logs to get the best lengths of timber. We've got to have access to some of that old-growth timber, otherwise we're gone.'

Recent reports have it that Gunns' proposed new pulp mill will be fed only plantation wood. Won't that mean no more sawlogs? 'Doesn't matter. Native woodchips will be exported,' Jaeger says. He means the trade will go on; the pulp mill is only one aspect of the timber industry. But I'm still confused about the underlying logic. Once, sawlogs were the sole reason for going into the forest. Today's contractors have pulpwood quotas, not sawlog quotas. The tall, straight, middle-aged sawlog trees are now seen as just a bonus. Forestry Tasmania estimates that the Upper Florentine contains 2000 tonnes of sawlogs, but it is only profitable to harvest them if the other 80 per cent of the forest is clear-felled and turned into woodchips.

Jaeger believes 99 per cent of people like nice timbers. 'But show them the process and they get worked up. It's like eating steak. They don't want to make the connection between the animal and the meal. But believe me, if we stop selling old-growth timber, the market will still want it. From Borneo, Indonesia, places with zero forestry code.' He is right. In the end, the responsibility lies with the consumer. But his comment also implies that the Australian market has taken a stand against bad timber practices overseas – whereas in fact, 400 million dollars' worth of illegally harvested timber from the likes of Indonesia is flushed through our market each year. It may be difficult to consume ethically when there are so many conflicting arguments, but do the bad habits of other countries' industries really justify complacency at home? Jaeger nods. 'I'm not saying everyone in the industry is squeaky-clean. It's good that questions have been asked. I've no problem with that. But we have changed. There's no need to shut us down. If we don't get access to that wood, we're goners.'

On my way out of his office, I ask to keep his drawing of the sawlog. Smiling, he pushes the bit of paper to me and, as an afterthought, flicks his business card over.

'They're like manure: if you don't spread 'em they don't do any good.'

*

'They mince-meat all the wood,' says Kevin Perkins when I try out the meat analogy on him – that consumers don't understand how trees are processed, that there's your prime cuts and your mutton. Perkins, one of Australia's top wood craftsmen, best known for designing the furniture in the prime ministerial suite in Parliament House, won't have a bar of it. He scoffs: 'Mince-meat. All of it.' He believes native forest should be managed species by species and used only for specialty timber.

Perkins' pieces embrace the unique characteristics of the Tasmanian forest – the mottled raindrop print of black sassy, the cat-fur patterning of tiger myrtle, the satiny ripple of Huon pine. But there is also a sadness in his work, an awareness that he is using the last pieces of a fading world. 'The chippers are going too fast for the crafts-makers and too quick for the trees to mature,' he says. 'There are ninety-odd under-storey species to the eucalypts, and they haven't a chance.'

Perkins' home is tucked in the southern valleys of Tasmania. The drive there is winding, with tall trees shedding bark on either side. My car hobbles around the paperclip turns and splatters like a tractor thanks to a little situation a few days earlier. I've taken to driving up the strange dirt tracks that splinter off from the highways, often after catching a glimpse of balding hills beyond the buffer of trees. Usually, I have to give up on the almost vertical rocky inclines and roll backwards to the bitumen. But every now and then I make it, my car hovering on the edge of an enormous clear-fell invisible from the highway. It feels right to see them, to not drive past. However, on my last attempt at rolling backwards downhill, I got stuck on a huge log. It was a crucial lesson in my manual-driving career. Revving, letting go of the handbrake, spinning the tyres, moving a couple of millimetres, quickly braking before rolling back, jumping out to wedge rocks behind the tyres – it took me an hour to get free and continue my blind man's roll back to the highway. The front passenger door stopped opening after that and the chassis began making a clunky sound, which thankfully disappeared with speed.

When I pull up, I'm outside a house that seems to tier into the hill. The Huon River flows somewhere down the valley – you can hear it – and peacocks whiz across the driveway, ducking their tiaras. Perkins barely says hello before taking a look at my side door. He disappears inside and returns with a butter knife and hammer. In minutes he has panel-beaten it into working order, laughing as I tell him how it happened. He has striking blue eyes and a rugged face, a rough tumble of grey hair and stubble. He takes me on a tour of the workshop and home he has built with his wife; it's like a beachcomber's paradise, except these are treasures from the forest, not the sea. There are sculpted birds and female torsos, a carved horse's head, perfumed shavings of wood, upside-down boats hanging from the roof, feathers, bones, fragile skulls and rocks.

In one piece of black-heart sassy – the name given to sassafras stained by rainwater after the tree's crown has broken off – he has used the watermarks to create landscapes; the ripple of tannin looks like seagulls pulling away from the grain. Using a trick from ancient Egypt, Perkins makes little wooden butterflies to keep the cracks and fault-lines knitted together. Roots, burls, knots and the creased skin of timber – all the things considered defects by sawmillers catch his eye.

Perkins left school when he was fourteen and did an apprenticeship as a joiner. After being conscripted to serve in the Vietnam War, he returned home to teach at a technical college. But when he met a wood sculptor from Canada who was intent on rescuing Tasmania's unique Huon pine from the flooding of Lake Pedder, Perkins was inspired. He started to work differently with timber, following its own lines and grain rather than the standard symmetry of carpentry.

As we talk, an unruly Dalmatian constantly nudges Perkins for attention, its tail thwacking the legs of the table. He points to a blue dog muzzle on a hook near the door.

'Present from Forestry,' he says. 'So she won't gobble up any 1080.' I ask him about Forestry's 'Island Specialty Timber' program, and he stands up. 'Let's go,' he says. 'You can see for yourself.' We drive back through the forested road and out onto the bitumen.

In Geeveston, there are wooden sculptures of ye olde people on the street corners, all shiny and orange as if they've been dipped in fake tan. Perkins drives into a muddied timber yard – Island Specialty Timbers. A local fella greets him warmly but warily, and hands us fluoro vests that we must wear if we want to walk around. It's not a big yard and the vests hardly seem necessary, but we shrug into them obediently. There are about five or six distinct sections, each stacked with tree trunks half-sunken in the wet.

'It's so lazy,' says Perkins. 'They don't even bother exporting the specialty timber. They just leave it here or in the north depot as promos for sustainable practice.'

He puts his hand on a piece of wood and flicks off the rotting bark so that I can smell its perfume.

'It's too young and it's rotting. They'll sell it for firewood.' He points over to another pile: 'Celery-top pine, one of the best boat-building resources in the world. Firewood.' There is a small shop in the yard selling polished brick-sized blocks of wood to turners and whittlers, or hobbyists looking to make the odd doorknob. Outside, Perkins picks up a piece of timber, a chunk of deep-red myrtle lying facedown in the mud.

'She can take this, yeah?' he asks the man on duty. The man shrugs, a bit overwhelmed by us. 'What's going to happen to it? Nothing? So she can take it,' Perkins affirms. He plonks it into the car. I'm not sure what I'm to do with it, but he is adamant I should have it.

When we get back into the car after giving back our hi-vis vests, Perkins spins the mud off the tyres and gets us out of there.

'For thirty years,' he says, hitting the steering wheel, 'for thirty years this island has missed out on so many opportunities while that industry is turning over huge profits. It's like they've gone mad – all they can see is woodchips. We're still the poorest state. We're such losers. We've got nothing for the woodchips. Our roads are potholed and dangerous. It's bloody-mindedness. Why spend all that money promoting the place if you're going to keep on ruining it?'

I ask him if he's considered getting a permit to go into the coupes afterwards, to salvage the leftover timber. He laughs.

'Salvage? It's a bloody tsunami after they go through. It's one company there to harvest the one thing – they don't give a stuff about anything other than woodchips. The harvester's machinery can't grab the smaller pieces, only big logs, and they won't go around trees they don't want, it's just smashed and pushed through. They say it's a renewable resource, but only if we fell and use a tree to make something that will last for as long as it takes for the tree to replace itself.'

Tasmania could have a sustainable native forestry industry if it capitalised on the island's unique and endemic timbers, Perkins believes. With more selective logging, and by placing a higher price on native timber products, the resource could last forever and turn a profit. This runs counter to the argument, made by Gunns and the state government, that the timber industry must 'value-add' with woodchips and pulp to remain viable.

'It would cost more to get the timber' this way, Perkins acknowledges, 'but it should cost more.' And the final product, he believes, would command more than the $12 to $15 a tonne currently paid to the state for its woodchips. Even floorboards aren't worthy of these native timbers, he believes; he is dismissive of the 'veneer mills,' which peel sheets from felled trees and claim to celebrate the unique island timbers. Perkins is thinking instead of deep-red myrtle tables, of boats made from 500-year-old wood that will float the salty seas for 500 more. This vision for Tasmania's forests is a much slower one than the current system, which relies on rapid turnover; 15,000 hectares of native forest are logged and burnt every year. Perkins' blueprint must scare modern loggers, many of whom have little desire to change how they work. He nods.

'It's much easier to smash everything, burn the rest for slash and electricity and pull out a couple of nice logs for promos.'

Perkins' criticism of current forestry practices is embarrassing to the government. Like the novelist Richard Flanagan, he is a source of much pride to the island and his opinions are not easy to dismiss. When he first openly opposed the state's forest practices, Bob Gordon, the managing director of Forestry Tasmania and former head of the Pulp Mill Taskforce, tersely reminded Perkins that he and the University of Tasmania Arts School have received millions of dollars from the state to 'enhance skills and marketing opportunities for designers.' Another critic of the industry laughs when I mention this scolding. John Young, who started the Wooden Boat School with his wife in 1992, is now a member of Timber Workers for Forests.

'Oh, we all get told off,' he says. 'Forestry Tasmania provided an annual scholarship at our school and then got angry when we criticised forestry for over-harvesting and burning future boat-building timber. But what good is a boat-building scholarship without having anything to build with? The scholarship looks good for them, but we'd rather have the resource than a scholarship.'

On our way back to Perkins' place, my muddied bits of hundred-year-old timber rolling around in the boot, he points out where his daughter found a dead wombat and heaved it into her car to bring home.

'The wombat was stiff with rigor mortis, but we could feel a live baby in her pouch.' Perkins had to use a chisel to open the sealed pouch and get the baby wombat out. For the next few months, they took it in turns to get up in the night to feed the little wombat, often falling asleep with the fuzzy brown creature in their lap. He takes me to see her when we get home. She's outside now and they're slowly introducing her to the bush. Perkins walks me past a shed where he is incubating chicken eggs, and past an old pale-blue Mercedes. Two peacocks are sitting in the front seat as if on a date, iridescent feathers shimmying across the red vinyl. Out of another old car pops a brown peahen, while gobblers and geese muddle around our feet.

We climb over a fence using a footstool and Perkins crouches in front of a lump of hay and blankets. He talks into a little dark hole and the hay slowly moves. A sleepy wombat emerges, bum first, groggily wondering what we want. Perkins is like a shot of coffee for her. He gees her up, rubbing her tummy, ruffling her head until she's running around playfully. By the time I get to hold her, she wants down for more wrestling. When we get up to leave she starts charging my ankles, headbutting me for play.

On my way back to Hobart, I decide to stop at Ranelagh. One of the activists told me about a logging family from way back, the Bennetts, who live here. These small operations work mostly out of sight in the forests, delivering their harvest to the mill door. Often family-run, passed from fathers to sons with the women doing the paperwork, these small businesses generally elicit more sympathy than the big corporations.

'They did their own blockade once,' the activist said when he told me about the Bennetts. 'On their own street.' A couple, two men, had opened eco-accommodation in the valley behind them and were having a launch party when the Bennetts, their employees and supporters closed off the road, stopping visitors (including Bob Brown) from getting through.

I only realise I've reached Ranelagh after I've driven through it, so I make a U-turn and head back to the smattering of houses. I'm looking for 'a T-intersection, house on the left, lemon tree out the back,' according to my activist friend. I take a punt, park and, feeling like a door-to-door salesman, walk up a driveway.

*

'Are you going to write a fairytale?'

Carol Bennett is leaning on the frame of their front door. Her husband, Tony, sitting on a bench under the carport, is telling her not to speak to me.

'We had this journalist,' Carol continues, 'she came out here, spoke to us, went out with the boys to the coupe and everything. Then she went back to Hobart and wrote a fairytale, like this is some bedtime story – there were two eagles in the forest, all lonely without the forest … It was a fairytale and they published it. Is that what you're going to do? Write a fairytale?'

Their dogs lean heavily against my legs. Again Tony tells her to be quiet. But neither of them can help it; their words fire up like an engine rolling over before the other one quickly kills it. Tony harrumphs from bench to stool and back to bench again.

'I'm fourth generation,' he says. 'My forefathers built this place. No one would be here if it wasn't for them. We built this place. My sons are fifth generation Tasmanian. If it weren't for us, the greenies wouldn't even know what forest is out there.' He sucks at the air around him. Carol takes over.

'She wrote a fairytale – so we've given up on talking to people like you.' Narrowing her eyes, she asks where I am from. I repeat that I'd like to leave my mobile number so they can call me if they're willing to have a chat – but I've got them worked up. I feel like a fox caught in a chicken-house. I tentatively ask about their miniature blockade.

'We didn't *block* anyone's way,' Carol says haughtily. 'We wore hi-vis vests and it was a safe protest, unlike them. People could drive through. We wanted them to know what it feels like to have their business protested against.' Tony presses his palms down on his knees and stands up. 'I'm not saying we don't need the greenies,' she continues. 'We do. We need them to keep an eye on the industry. But when they stop us from going to work in the bush, that's just not on.'

Tony breathes out.

'All right, Carol, we better hold it there. Better not say anymore.' He looks above my head and points at the hill behind us. 'That's my fire trail up there. I made that track back in 1968 and now it's Bush Retreat's walking track. The two greenies up there think they own that track. Their flyers say it's all virgin forest round here. Pig's arse – I logged it in '66.' He harrumphs back down onto the stool. Carol looks at him, then back at me.

'The media – they report everything wrong. We write letters and they don't run them. We let them know we're donating money to the Royal Children's Hospital and they don't turn up. Or when our boys went to Victoria to help out in the bushfires, the media never wrote that up.'

I look down at the dogs headbutting my legs. How do I explain that the media don't like good news, and that their attempts to generate positive coverage would seem like crude public relations to most journalists?

'You're probably not media-savvy enough ...' I start to say, trailing off as Carol keeps talking.

'And a defibrillator, we bought one for the local ambulance, not one word.'

Tony gets up, shoos the dogs away from me and starts walking down the drive.

'That's enough now, Carol. Enough.'

Carol tells me Tony started out cutting timber for crates in the apple industry. Today he, she and their two sons run T.P. Bennett & Sons.

'I mean, it's a forest,' Tony says, turning around and walking back to us. 'It's a living thing – you can't save that. They're meant to die. And underneath the forest there is more forest, I found … go inside, Carol, and get that bit of petrified wood. She'll show you.'

Carol goes inside and again Tony waves the dogs away. They slink behind the car, but as soon as I hang my hands down they're back, nudging their heads into my palms, tongues on my wrists. Carol comes out holding a sandy-coloured slab and makes me hold it.

'Feel how heavy it is? That's petrified wood. Tony found it under a waterbed.' I let it drop a little to acknowledge its weight. She takes it back carefully and goes back inside. Tony looks at me.

'See?'

There was wood and there will always be wood, I interpret.

He walks back down the drive and gets into a little jeep. The dogs jump on the back. I ask him and Carol what they think about the growing population of young activists in Huonville. Tony just snorts, his hands tightening on the steering wheel.

'Believe it or not,' says Carol, 'I'm intimidated when I walk down the street and there's a big group of them.' I can see her shrinking from them, holding shopping bags close. Not smiling or saying hello; she no longer knows all the faces in town. Tony starts the jeep.

'We better stop there, Carol. Look in on the cooking there. We better stop there.' He turns the wheel and disappears up the hill, the dogs barking happily – one jumps off and races him. Carol keeps talking.

'It's not so much like this anymore, but back in the '80s when the heat was really on us, you couldn't even trust your kids to their school teachers. I remember one time my daughter coming home and crying. She asked if Daddy was a murderer.' She shakes her head. 'Imagine your daughter being accused of that. And these ferals, they wonder why there's hatred. In the morning we'll get a call from one of our men around 4 a.m. and he'll say, "They're here, missus." It's just him in the cabin and about twenty or so ferals surrounding the truck outside. I mean, it's intimidating.'

Carol and I are still in the driveway when Tony comes back over the hill. He parks the jeep, yells at the dogs, tells Carol to quit yapping and says goodbye to me. Firmly.

BURNING OFF

Michael Woods lives in the township of Triabunna, a fishing and woodchip port halfway up the east coast. A timber contractor, he has agreed to speak with me after a series of phone calls to various timber groups. His work yard is behind a petrol station and the small office is a portable building on temporary stumps. Hanging on its thin walls are framed photos of excavating machinery and trucks. Woods employs about sixteen men and works coupes around Snow Hill, typical dry Australian bush where cicadas sing and the last light is orange and dusty with insects. In the evenings the men regroup at the work yard, shimmying over the trucks and investigating any clunks or clangs. When I arrive, it's just before noon. No one is at the yard but Woods and his secretary. I shake his hand. He is, like many of the men I meet, well fed, his skin a ruddy mix of suntan and sunburn. He drives me to see where his crews are working.

As we travel up the coast, the turquoise water and white sand glinting, he points out his favourite place, Wineglass Bay. 'That place is the jewel in Tasmania's crown, I reckon.' Crayfish season has opened and amateur fishermen have a month before the commercial fishers start pulling their craypots in. 'I had my first one last night,' he says contentedly. On the other side of us, it is dry paddock after dry paddock, dead trees jutting out of the hard soil like shrunken knobs of ginger. The drying out of Tasmania's east and midlands was noticed more than 150 years ago by a British settler who, after noting that trees spared during the rapid clearing and grazing were dying, wrote, 'It was as if a locust has passed over the land.'

Recently the state government announced a billion-dollar scheme to siphon water from the west and north to reverse the effects of the drought that has seeped eastwards from the midlands. Tasmania will then become, the premier predicted, Australia's 'food bowl.' Woods radios his neighbours as we pass their farmhouses, explaining their situation to me between calls. Most are praying for rain. I ask if he thinks it's because of global warming.

'Nah, it's just cyclical, this stuff,' he replies.

Woods gets back on the radio to find out how many millimetres of rain have fallen this year. 'Less than half this time last year,' a voice crackles into the car. Woods sighs and turns inland, entering the dry eucalypt forest of the eastern tiers. Bunny-hopping along a dirt road, he stops abruptly every few minutes to point at the bush. 'There's regrowth,' he says, 'and there, and there.' Pointing to a tall stag tree poking up out of the bush, he says, 'A wedge-tailed eagle's nest was up there. We had been driving past it with our trucks the whole time and never bothered it.'

Woods quizzes me repeatedly: 'How old do you think that bush is?' 'Thick as the hair on a dog's back, hey?' 'You'd never know it's been logged before, eh?'

I agree the regrowth is thick – but is it too thick, I ask. It was surprisingly easy to walk through the old-growth forest in the Florentine, but the trees here are packed in tight like a box of matches. Woods nods. He explains that the bush is 'suppressed' and his workers will thin the regrowth out, injecting about a third of the trees with Roundup, a weed killer. Standing on a steep slope, he describes how a recently introduced method of cable logging will lessen the footprint of machinery. He explains that the trees will be felled by hand, then connected to an overhanging cable and dragged to a landing. I ask if it is even possible to use machinery on such a sharp edge.

'No,' he answers.

At the edge of a massive clearing, I ask Woods if this coupe will now be burnt. Every autumn in Tasmania, over 200 forestry burns are planned. The horizon is regularly flanked with mushroom clouds of smoke and the island's famously crisp cornflower-blue skies become thick with ash. One of the most controversial practices within the industry, this burning off is what the timber industry calls 'regeneration.' Forest ecologists call it a transformation at best, as only commercially viable forest is allowed to grow back. Asthmatics call it hell. Woods nods and concedes it will be burnt, but is ready with a defence. 'The Abos used fire to regenerate the land – it's a proven method,' he tells me. When he sees the cynicism flash across my face, Woods thinks I doubt this. 'They did,' he says. 'They burnt this whole place.'

I'm not sure how to respond. It is true, the Aborigines did tend to the land with fire; early British settlers observed that the island was regularly wreathed in smoke. Farmers wrote home excitedly about the vast and initially fertile plains, declaring Van Diemen's Land an agricultural Eden. The soil's fertility is now known to be largely a result of the Palawa people's slow-burning regimes. Today, many Indigenous communities elsewhere in Australia still have a special relationship with fire. I have seen teenagers outside of Alice Springs handle fire as if it were twine, drawing it back and forth with their fingers and sticks, dabbing it onto grass and then sweeping it around their backs like a cape. But I'm uneasy with Woods' declaration. In Tasmania, Indigenous culture is seldom invoked unless to defend the forest industry's clear-fell and burn regimes.

This year activists let off flares outside the Forestry Tasmania head office in Hobart, forcing the director to admit a 'slip-up' in a series of recent burns in the Huon region that made ordinary activities near impossible for many residents. But despite issuing the 'sincere apology,' Bob Gordon reasserted that 'there is no alternative' to the current practice of burning coupes after logging to spur the regeneration of eucalypts. Helicopters are known to swoop over cleared coupes and drop ping-pong-ball-like incendiaries, igniting piles of immature timber, branches and logs that have been left to dry for a summer. Hand-held drip torches are also used to release a gluey substance that lights up like a fox's tail. The substance dropped from the sky is similar in effect to napalm, hence the mushroom clouds, but the timber industry – understandably – prefers to call it 'petrol gel.' This sounds to me, the first time I hear it, like a hair product. In truth it is a form of jellied petroleum, which is the same substance found in napalm *and* in Vaseline.

'If anything, we're the conservationists,' Woods continues. 'The eucalypts need to be burnt, otherwise the other species will take over. They could become extinct if it weren't for us. The rainforest would take over.' Somewhere in this thinking there's a seed of truth. For a time, botanists believed the eucalypts were unique to Australia and that the rainforest must have snuck over from Asia, taking over our tall forests. But fossils have since shown the eucalypts to be a rather recent addition to the landscape – some 34 million years ago – while even older fossils reveal that rainforests once grew throughout the continent, including in now arid places such as Lake Eyre.

'The Abos—' Woods starts again, but I cut him off. I want him to stop saying 'the Abos,' and to stop using them to defend his business.

'Do you know any Tasmanian Aborigines?' I ask politely.

'No,' he admits.

*

The truth about Forestry's burning off is that it is the best way to grow back a predominately commercial forest to be re-harvested on a 35- to 90-year rotation – the time period set for logged coupes to regenerate before another harvest. An aerial burn over the piles of leftover wood in a logged coupe burns away the moist, organic humus on the forest floor, speeding up the decay of the debris and any surviving plants, and thereby nourishing the dirt much more quickly. The exposed minerals allow specially selected species such as commercial eucalypts to grow back quickly. But the soil is worked hard in the process. How many times can it be burnt like this before it starts to fail?

Older bushmen who used to work in portable sawmills out in the forest recall how the sawdust they left in their wake was lapped up by the forest; the sawdust kept the soil wet as seedlings pushed through. But applying the language of biology and ecology to mass cropping and farming is misleading. In particular it is disingenuous of Woods to compare today's industry practices to Indigenous burn-offs. The Palawa people didn't clear the land and then 'petrol gel' the exposed soil with helicopters. And while there is evidence that Indigenous burnings may have altered the structure of certain forests over a period of 10,000 years, Tasmania's landscape has changed more rapidly in the past 200 years of British settlement.

What Australians call 'deforestation' in Indonesia, Papua New Guinea and the like – that is, the clearing of native forests in order to plant crops – is called 'afforestation' in Tasmania, because often they 'let the forest grow back' and therefore, according to the timber industry, shouldn't have to account for their carbon emissions if an emissions trading scheme is implemented. But 'letting the forest grow back' is misleading, especially when it will initially involve the clearance of a forest older than its intended rotation, for example a 500-year old forest that is cleared and then logged every forty-odd years. A regrowth forest, as managed by Forestry Tasmania, is more likely to consist of a crop of eucalyptus trees all of the same age and type, harvested for uniform pulp and sawlogs. Sure, the trees aren't in rows, and some other species may force their way back for the second rotation, but all in all, it is grown with commerce in mind. Yes, as the timber workers keep telling me, the trees will grow back. But the forest that was there before is gone forever.

I suspect the reason my phone calls all seemed to lead me to Michael Woods and the east coast was to get 'the journalist from the mainland' away from the romantic wet forests with their towering ferns and rising mist, and into typical Australian bush – the place of battlers, hard work and seemingly 'un-contentious' forest. But ironically, I am now close to a very different and complex epicentre of the forestry's operations.

In the east and north-east, local conservationists say the dry eucalypt forests and grasslands are over-logged and converted into plantations. It is even suggested by some that these areas are targeted because of their location. Being so close to the woodchip mills and ports, their proximity undoubtedly cheapens the transportation costs. But because this type of vegetation mostly occurs on private land, conservationists say it is even harder to advocate for its protection. However, with considerably fewer sawlogs within these drier areas, it is curious that they are being logged at all.

'I can't hold my head up this year,' says Woods when I ask how much of his crew's haul from these forests are chipped in comparison to sawlogs, 'but two years ago, we delivered about 42 per cent saw-logs.' A local conservationist who lives further north, just outside St Helens, Todd Dudley, doubts even this already low figure. Dudley believes the logging in these regions is almost purely for woodchips. 'I've seen forest logging plans for coupes that have extracted 98 per cent woodchips and only 2 per cent sawlogs.' But, he adds, in the past three years the volumes have been censored on the forest plans. 'We can still demand to see them, but they black out the volume quotas with markers. They say it is because of "commercial-in-confidence."' He laughs. 'I think they're just embarrassed. After all this time of saying woodchips are just the residue, it's the by-product, you can't have numbers like that floating around.'

Our next stop is at a section of forest logged eight years ago, next to untouched forest. Michael Woods stops and beckons at me to get out of the car. Standing on the track, he points. 'It's just the same, see? Both sides have lots of trees.' Not wanting to be nit-picky, I nod. But there is shade on one side. It is cooler. This is the side that hasn't been logged. Internally, I sigh. I am on a journey through selective truths.

*

Tasmania's forest battles are fought largely in the arena of images and semantics. Green campaigners work on strengthening the public's connection to the wilderness, while the timber industry works on disconnecting people from these 'unmanaged' forests. The Wilderness Society's aerial shots of scarred charcoal swathes of land are succeeded by Forestry Tasmania's counter-images of the 'tapestry' of regrowth.

The books, DVDs, brochures and pamphlets I receive from FT are a journalist's nightmare. I find myself constantly trying to decipher new words. Nature needs 'disturbance,' logging is 'harvesting,' deforestation is 'afforestation,' burning woodchips for electricity is a form of 'bio-fuel' or 'renewable energy.' Woodchips are 'feedstock,' while the non-commercial attributes of a forest are 'non-wood values.' Today, Forestry Tasmania claims clear-felling is kept to a minimum. Instead, loggers engage in 'aggregated variable retention,' whereby small circles of the forest are left intact; the wilderness can creep out again once the chainsaws, trucks and excavators retreat. These clumps, sometimes the size of roundabouts, are often scorched, windblown and falling over. From an aeroplane, they must look like hastily drawn crop circles.

The definition of 'old-growth' in particular is increasingly murky. In forests such as the southern Weld Valley and Middle Huon, there are trees well over 300 years old alongside younger trees that sprouted up after wildfires in 1904 and again thirty years later. As a result, the trees are of various ages, which in itself differentiates them from plantations. But according to the timber industry, this means the forest has been 'disturbed' and can be classed as 'regrowth,' which, linguistically, makes it seem less sacred. It's easy to see how activists get cynical: you could piss in an ancient forest and have its status change as a result. The definition of a rainforest is also malleable. In Tasmania, if a rainforest includes more than 5 per cent eucalypts, it is no longer a rainforest – which means parts of the Upper Florentine, despite its mature rainforest under-storey, is a 'wet eucalyptus forest.'

Standing in a cleared coupe, Michael Woods plucks a eucalypt seedpod off the ground and shakes it over my hand. Seeds like tiny flecks of chocolate fall into my palm. He explains that forestry officers use guns to shoot seedpods down from eucalypts with tall, straight trunks.

'[Trees] that don't have any defects will generally make more of the same. When it's time to sow, they take a sack of these' – he points to the flecks – 'and drop them out of a helicopter.' We look up at the sky. I imagine seeds already shucked from their pods, thousands of them falling, getting in our hair and clothes. Recently Woods went further into debt to purchase million-dollar cable-logging equipment to win a tender to extract wood from a nearby steep incline. He has a lot to lose out here.

*

It is difficult to describe a clear-fell. It is like a sharp intake of breath. What you were looking at a second ago is suddenly gone, as though an enormous cartoon trapdoor has opened beneath it. And after it is burnt, what remains is like the remnants of a campfire on an enormous scale: a black pool of charcoal and ash fluttering around like moths. On a scouting mission with the Florentine blockaders I was taken to see coupe SX10F, a section of forest in the Styx. It has been cleared but not yet burnt. The four of us tread down the hill. It is quiet except for the snapping of twigs underfoot. The forest has been kneecapped; skinny trees bend over, cracked at the waist. Our hands absently touch collapsed man-ferns and a few pokey shrubs. The odd spindly survivor stands up, having dodged the metal claw and rubber tread of the bulldozers, but next year it will be burnt. Miranda tells me this coupe was felled last winter.

'There weren't enough of us to go scouting. My heart was breaking. I just knew that while the forest we were in was standing, coupes were falling all around us.'

To Ula, this coupe taught her more than she ever bargained for.

'I'm usually okay, but seeing 10F killed me,' she says. She believes this particular coupe was cleared out of spite. 'It was a revenge operation. At all our meetings with Forestry, I kept pushing for them to keep the chainsaws out of it. They knew how much we cared about it.' Ula learnt a hard lesson in forest activism. Never specify *exactly* what you want protected 'because they'll probably be chalking it up on the priority chopping block before the meeting is even over. If I wasn't cynical before, I am now,' she says.

Looking at the three women beside me as we stand in this clearfall, I find that I can't match their sadness and wonder if they're setting up their hearts to be broken by visiting and falling in love with places almost certainly doomed. They live in them, learning the habits of particular animals and the songs of particular birds.

'We need to start taking our kids to places that aren't going to be destroyed,' one activist and mother of two told me earlier. She discussed it with her partner after their little girl asked when they were going camping in 15F again. 'I said, well, they're logging it, and the smile just fell right off her face. Like something had died. Which it had, I guess, but I hadn't thought of how that might affect our children. It had just become so normal for us.'

'One-hundred-and-seventy years of history destroyed in an afternoon,' said a reporter on the Hobart evening news when the Myer department store caught fire in 2007. The historical facade with its ornate macramé decoration and bell-shaped windows looked like an old film set as flames poured out its windows and fire fighters doused it with huge ropes of water. People gathered to watch the fire burn late into the night. Some wept while others held up their mobile phones, taking pictures. Today the burnt shell is referred to as 'the hole in Hobart's heart.' It remains untouched. In town one afternoon I decide to visit the blank spot on Liverpool

Street where Myer once stood. Is it possible to feel the absence of something you never felt the presence of? And is it the same with Tasmania's forests – can you miss something you've never seen?

Standing in the clear-fell, I suspect I'm no different from people all around the world, standing in places long gone, trying to feel the shape of a ghost, trying to feel something. In describing their sense of loss, the environmental movement can often seem melodramatic. Evoking napalm, Hiroshima and the holocaust to describe logging is manipulative, although it can be hard to find definitional error with such words. Holocaust comes from Greek, *holos* meaning 'completely' and *kaustos* 'burnt.' To burn away, leaving no trace of what was previously there.

In SX10F I climb up onto the stump of a huge tree. What's left is mostly hollow. There's enough room for a tea party in there. Bridget climbs up with me. She points at its pithy insides and in a big blokey voice mimics a logger: 'It's rotten up the inside, love! Dead anyway.'

We laugh. The sound echoes in the vast empty space.

JOBS

Loggers didn't always defend the practices of the wood-chipping industry. When the woodchippers started clear-felling in the 1970s, the original bushmen put up a fight, accusing these new visitors of taking away their future yield and tearing up the forest floor. In a clear-fell, every tree, shrub and fern is felled. Then the trunks, if of use, are carted away for processing, leaving the rest as 'slash' to dry and burn. Thanks to paper mills such as ANM starting up in the 1940s and '50s, older bush workers grew used to changing work practices. But nothing prepared them for the speed with which the chippers went through the forest. Soon loggers were instructed to fell all the sawlogs in one swoop and let the chippers take the rest, flattening the forest as they went and leaving no future sawlogs to come back for. Fights broke out between the two operations, but the chippers had immediate demand and fast financial turnover, and soon took priority. The tension between the different forest operations was dealt with by merging the original loggers with the wood-chipping crews. Bush knowledge that had been handed down from generation to generation was lost now that the men were employed to cut, drag and drive – no more, no less. A supervisor would make the final decisions about where the wood would go. An uneasy peace was restored.

Today, individual logging crews compete to fill supply quotas. I spend a day going through the business directory phoning loggers and haulers, and for good measure, checking in with 'earth excavators' too. Many numbers ring out or are disconnected, and a few families are trying to sell their businesses, unfinished contracts included. Of those who do speak to me, most want to remain off the record when criticising the industry. Voicing even the tiniest criticism makes them nervous. North Forest Products, later bought by Gunns Limited, was the first company on the island to outsource its logging operations. It didn't take long for the entire industry to follow suit. By transferring the costs and risks to small businesses, much of the headache was removed while profits could continue to soar. Not that the logging contractors saw it that way at first; this new independence seemed to offer them a share of the good times when timber demand was high. The downside only began to reveal itself when Gunns suddenly increased its asset base in 2001, from $143 million to a whopping $703 million, after acquiring both North Forest and Boral's Tasmanian forestry and woodchipping operations. Gunns was now the biggest hardwood-chipper in the southern hemisphere, controlling more than 85 per cent of the island's forestry operations. At the mercy of one company, the bush crews found themselves powerless during the tendering process. 'They treat us like dickheads,' one contractor says to me. 'As if we don't know what we're worth.'

'Well, we don't, do we?' Kevin McCulloch says when I repeat this to him. Leaning back in a plastic chair in Hobart's Salamanca Place, he shrugs. 'We under-quote ourselves and even then there's always someone who will do it for cheaper, and cut corners no doubt.' McCulloch is the only timber worker I meet who is completely at ease in what is probably Tasmania's trendiest district, a busy waterfront stretch of art galleries and restaurants. Wearing shorts and a T-shirt, he is friendly, good-looking and considered. He tells me about being a house-dad for two years before taking over his father's timber business. 'It was great being a house-dad. I loved it, but the money started to run out. So I had to go back into the workforce. I was weighing up between going into construction or back to forestry.' He pauses. 'I made the wrong choice, didn't I?' I ask him why he chose it. 'I don't know. I guess once it's in your blood, it stays there. There's a real freedom in it as well, being out in the bush.'

McCulloch runs two forest crews. 'One crew thins, which is basically pulling out the small wood in regrowth. They're a good honest team – they look after and maintain the equipment, do their jobs well and with pride. The new second team is a clear-felling operation ...' He pauses. 'They're, well, there's a real lack of care-factor. I'm planning on moving them into plantations.'

After much to-ing and fro-ing, I begin to get a sense of how the contractors work. The principal in the contract (usually Forestry Tasmania or Gunns) will estimate how much wood is in a coupe and put the job out to tender by competing bush crews. To win a contract, loggers must not only have the right equipment (which requires hundreds of thousands of dollars in bank loans); they must also undercut their competitors. The successful team will receive an agreed price to extract the wood and haul it to the mill door; the mill then pays the principal an agreed price for the wood.

'We're key to the industry and we're treated like shit,' says McCulloch. He says contractors can only just cover their costs by fulfilling their contracts – but Gunns' contracts now include a clause allowing the company to reduce the quantity of timber they will commit to pay for if the market changes. In 2006 the Supreme

Court ruled in favour of a logger, Noel Jackman, who took Gunns to court, claiming that the company had no right to terminate a five-year contract with his business when it still had more than three years to run. It was a victory for the loggers, but a temporary one. The following year, the state government altered the Forestry Fair Contract Code, allowing timber companies to cancel, reduce or shorten their contracts if demand changed.

McCulloch is a core member of the Tasmanian Forest Contractors Association, a group set up to give loggers some leverage in the industry. They have been trying to recover some of their bargaining power. 'But they refuse to deal with us as a collective; they'll only speak to us one on one,' McCulloch says. 'On top of that, we're responsible for insurance, wages, workers' comp, industry training and whatever else.' When I say it sounds as though they are trying to form a union, a luxury they relinquished when they went out on their own, McCulloch laughs. 'Pretty silly, hey? Not the smartest place to end up.'

In 2004 during the federal election, bizarre images flooded the media of timber workers in Launceston cheering the then prime minister, John Howard, a staunch anti-unionist. Howard waved like the Queen to a sea of men in fluoro orange vests and polar fleeces. Forestry union spokesmen had psyched up the crowd, which included timber workers and contractors, for Howard's entrance. If they wanted a job after the election, the union leaders told them, then 'Cheer, goddamnit, vote Liberal, look after ya mates.' A kind of fervour overtook the throng. Unions on the mainland recoiled in horror at the images, warning that John Howard was no friend of the workers and that the Coalition forestry policy was a 'con job.' But the cheering only got louder: 'You're a fucking legend, Howard!' one timber worker was heard to shout.

McCulloch squirms uncomfortably when I ask him if he was one of those cheering for Howard in 2004.

'Yeah, I was for him. He promised us a job come Monday.'

'But not Tuesday?'

'Yeah. Not Tuesday. You think we're pretty stupid, don't you?'

*

In Snow Hill, Michael Woods offers to get his 'bush boss' to fell a few trees for me. We are on a private property; the owners sell their timber to Gunns. Without thinking, I panic.

'No!'

Woods looks at me oddly.

'I mean,' I say more slowly, 'not unless you're going to fall them anyway.'

He laughs.

'Well, of course we're meant to fall them. That's why we're here.'

The 'bush boss' climbs inside a yellow machine, similar to a bulldozer but with a blade extending from its arm, and drives to a section of eucalypts. It takes about eight minutes. Three huge trees, each about 15 to 20 metres high, fall on their sides. The sound reminds me of the time, when I was in primary school, one of the other kids fell off the monkey bars. There was a sickening crunch as he fell on his arm – the sound of a bone breaking. Except when a tree hits the ground, there is also a huge thump and the earth jumps under your feet. Activists in tree-sits in the tropical forests of north Queensland say they can feel their own tree quiver when another is felled, even if it is some twenty metres away, because of the deep and intricate root systems.

On the other side of the coupe another man juggles a log with the mechanical claw of his excavator, dropping and pinching it until it splits in two. It is like a huge version of Skill Tester, the arcade game where you press buttons to manipulate a little metal claw, with which you try to pick up stuffed toys and key-rings. I get the feeling these yellow machines are a bigger threat to timber jobs than any greenie.

It is difficult to get a definitive sense of how many people are employed directly and indirectly by the timber industry, especially as employment figures make no distinction between plantations and native forest logging, two fundamentally different sectors. In 2005 the Australian Bureau of Statistics stated that 3500 people were employed in timber manufacturing. The Co-operative Research Centre for Forestry, based in Hobart, estimated the forestry and wood-products industry employed around 6300 people in 2005–06. In 2004 Timber Workers for Forests (TWFF) estimated that no more than 8000 Tasmanians were employed in the industry, with just 300 men working in the most controversial forests. A contractor, who asks to be anonymous in order to keep his job, scoffs when I mention this figure. 'Try about 150 now. And it's still too many men. But the more of us competing for contracts, the cheaper we are.' Given Tasmania's population of half a million, of whom 222,000 people are working, even the inexact numbers suggest that forestry accounts for no more than 3 per cent of the island's workforce.

Most loggers blame the lack of work on the environmental movement and the creation of protected reserves, but jobs in rural areas have been disappearing since the late '60s, even as the woodchipping industry took off and the volume of wood produced increased by almost 40 per cent. It didn't matter that demand for woodchip soared as Japan's paper industry rapidly expanded – the jobs continued to disappear.

At the moment there are eight jobs directly hinging on the future of logging in the Upper Florentine. In old-growth native forests, men work in teams of four, or sometimes eight, but rarely all at the same time. It is a lonely day's work, as opposed to the good old days, when it took up to ten men to fell and cut one tree. While machinery has made the work safer, it has also made many jobs obsolete. Machines don't have families to feed, and don't ask to take holidays or collect compo. They've done far more than the greenies to put timber workers out of work. And yet the forestry conflict continues to be framed as a choice between jobs and the environment.

*

Both Michael Woods and Tony Jaeger believe that if logging were to stop in contentious, high-conservation forests such as in the Styx, the Upper Florentine and the Weld, the entire native-timber industry would shut down.

But in the 1920s, according to Judith Adjani in her book *The Forest Wars*, that was part of the plan. An economist at the Australian National University, Adjani has twenty-five years' experience and understanding of the forestry industry. She describes early British settlement in Australia, when the native eucalypt hardwood timbers were considered difficult and unwieldy in comparison to the North American and Baltic softwoods, which could be quickly sawn through for boards. The eucalypts required patience. Fresh off the saw, native timber could be used for railway sleepers or house frames, but for other uses it had to be dried for years. In the late '20s, a prominent group of national foresters decided the only way the Australian timber industry could survive was if it planted softwood. Despite its name, this is not always a softer type of wood;

the difference between 'hardwood' and 'softwood' is simply in the seed – the former produces seeds with a covering, like an apple seed, almond or gum-nut, while the latter drops naked seeds. Some fifty years later, in 1967, the foresters finally got their 'softwood solution' and $450 million in federal loans.

However when the foresters said softwood plantations would 'replace' native forests, they meant it literally. Having spent the nineteenth century battling the colonists' tendency to 'deforest' everything for farmland, the foresters now began their own deforestation program, converting native forest to softwood plantations. By the 1980s, according to the softwood solution, it should have been time to let the native-wood industry run down and plantation forestry take over. Instead, state and federal governments submitted to lobbying pressure from industry and native woodchipping escalated. Adjani identifies three reasons for this: the timidity of the plantation sector in demanding recognition of their role; the 'extraordinary profit performance' of the woodchip industry; and the alignment of the forestry union with the native forest sector.

In 1996 the federal government produced yet another plantation initiative, the '2020 Vision,' this time rolling out native hardwood plantations across the country. Environment groups were pushing for plantations to replace native forests for wood and fibre, and many saw the tree farms as an end to logging in native forests. But rather than fading away, logging of native forests exploded.

For companies such as Gunns and Forestry Tasmania, what was meant to be a replacement for native forestry – plantations – simply became an expansion of their business. It's like a chicken farmer selling both free-range *and* caged eggs; his free-range prices are cheaper than those of his competitors who sell only free-range because they are offset by profits from the caged eggs.

The big companies now reap profits from both native and plantation products. Australia's plantations currently yield more than 80 per cent of the nation's manufactured wood products, but native forests dominate the woodchip sector. Native woodchips sell for around $12 a tonne, while plantation woodchips, to be profitable, cannot be sold for less than $30 to $35 per tonne. For the most part, the two sectors are selling to the same woodchip markets. If native forests were producing unique, irreplaceable timber, the case for logging them might be clearer. But critics of the industry believe the forests' unique product, its specialty timbers, and its useful product, sawn hardwood, are being over-cut, cut prematurely and replaced with the fast-growing eucalypt pulp trees.

The 2020 Vision lifted protections on agricultural land, native forest and rural residential land and within a decade in Tasmania alone, 100,000 hectares of native forest on public and private land was converted into plantations. In rural towns, the change was dramatic. Some were reduced to nothing but a road sign. This second plantation roll-out has created a whole new arena for the forest conflict, with farmers, local landowners and 'non-greenie' types standing alongside environmentalists and greenies in opposition to forestry. Locals and farmers claim that thanks to the thirsty trees, their springs are drying up and creeks ebbing to a trickle. Complaints about aerial spraying are constant, with people claiming to have been drenched in chemicals and children called in from the schoolyard as a pesticide drift soaks through their T-shirts. The health of rivers, bores and water-tanks is now uncertain. The animals that kept the grass low are baited or shot and local landmarks are cleared, replaced by repetitive rows of the same tree. Some residents say they feel like tiny spit roasts in amongst hectares of kindling.

*

'By 2050,' Michael Woods, the logging contractor, declared at our meeting, 'there will be more trees in Tasmania than before white man arrived.' As I drive inland through the north-east, I believe him. A patchwork quilt of tree farms has been laid over the land. From dusty lilac, pine green to toffee red, squares of trees are sewn close to the road and up over the hills. If only it were as easy as this, I think, and this comforting blanket of trees could stop logging in contentious forests. I drive past 'Fern-a-mania' before slamming on the brakes and reversing alongside it to get a closer look. Piles of tree ferns from cleared forests lie on their sides in the yard.

When I arrive at Terry Rousell's place in the Upper North Esk, I spy him and another man in the paddock rolling up fencing. Both are wearing blue flannel shirts. Climbing over the palings, I head over, finding a little blonde girl wearing a velvet dress on my way. Together she and I walk to the men, who have stopped work and are watching a helicopter buzz back and forth like a blowfly along the slope of the valley.

'Nice timing,' Terry says, shaking my hand. 'If you want some first-hand experience, go stand over there' – he points to the slope – 'and get a nice spray of pesticide.'

The three of us sit and watch, while the little girl – Terry's daughter – walks around us, swatting the march flies off our legs and arms.

'He wanted to park his chopper here once,' says Rousell. 'I told him he was dreaming. He's too close as it is. He wasn't surprised when I said no. "I can't blame ya," he said.'

Between the chopper and us is the North Esk River. It is hard to tell how far away it is, maybe 200 metres from us to the river and then 100 metres up the slope. But in a gully the wind seems to change depending on where you stand, and it's the wind and rain that bring the chemicals close. A local I met outside Deloraine, an inland country town, told me he and three friends were drenched in aerial spray one afternoon while having lunch at his friend's property.

'It was so close that we got the camera out to take photos and then we just got covered in it. It left this strange metallic taste in our mouths.' The next day, he said, all of them felt strange but didn't think much of it. 'Everything looked foggy and our eyes hurt. It was so misty I ran my car off the road.' It wasn't until they spoke on the phone that they realised their slightly off-beat health might have been caused by the spray. They sent the photos to the Forest Practices Board, but 'we got the usual run-around.'

It's strange how instinctive it is to be cynical about anecdotes like this. Perhaps because chemicals are invisible, and the chain of cause and effect so murky, it's tempting to think people are making such stories up. I'd be even more cynical if I hadn't once swum in an accidentally over-chlorinated backyard pool. We didn't notice it immediately. My brother had opened his eyes underwater, swimming around grabbing at my legs. Within an hour of getting out, he started howling and rubbing at his eyes. When we realised the pool was way over the normal level of chlorine, we assumed I'd managed to escape its effects. It wasn't until later, while we waited for my brother to see a doctor, that I said absently, 'Is there a house-fire around here? This smoke is intense.' There was no fire, no smoke. Sitting on the hill with Terry and his daughter, the helicopter so close, I don't find it hard to believe people have been caught in pesticide drifts or had their drinking water contaminated.

The spray runs out beneath the chopper's belly and it flies over to Sunset Ridge to refill. Behind us is Terry's old sawmill.

'My dad and his brother were bush men, and they taught me.

They used to go out with an axe and cross-cuts.' On three sides of the Rousell family home are plantations and out the back is state forest, where Terry's father and uncle used to cut.

'Dad reckons they could have logged that block forever – doing a sweep, running up the wood, and then another sweep … forever. Now? They're just balding the mountain,' he says. Back then, saw-millers had permits to collect a quota of sawlogs from the forest nearby. But since the rise of woodchipping and anonymity of logging crews, sawmillers were being allocated wood from all over the island, with chippers influencing which areas were cut. The sense of responsibility Terry's father and uncle felt towards the forest diminished as it was cordoned into numbered coupes to be clear-felled.

'We had a contract with Forestry Tas for 1000 cubic metres of blackwood, which they renewed, but we only ended up getting 600 cubic metres the second time round. We'd pay for a certain amount of sawlogs to be delivered and we'd get category-two logs when we had paid for category one, or worse, logs that you couldn't get a cut out of. The drivers would tell you there weren't better logs available.' After sawmillers started getting angry at the log-truck drivers, one harangued driver started handing out the phone number of John Gay, the executive chairman of Gunns. Rousell called him up.

'We had an amicable chat at the start, but when he asked me how he could get people to stop picking on Gunns, I was mad. Picking on them? There used to be twenty sawmillers around here. Now there is one, and he cuts for Gunns. He hung up on me.' When the time came for Forestry Tasmania to renew Rousell's contract, they said there was no more blackwood left. 'But there was,' he says. 'It was just going to Gunns' sawmills instead.' His contract wasn't renewed. 'We couldn't beat 'em,' he recalls; the bigger timber companies were too strong, he says, buying out the small mills and redirecting their timber to their mills. 'If you complained about the quality of logs you were getting whilst you saw perfectly good logs heading to the chipper, they'd cut you off.'

In 2000, Rousell shut down his sawmill. 'I'll never cut another tree down. I can't justify it now. Not now.' The valley in which he, his wife Megan and their daughter live is a shadow of its former self. Two schools, a post office and the small sawmills have shut down. Many neighbours have left. There is a sense in these tiny towns, dotted like fading constellations along the highways, that the contractors, plantation owners and remaining sawmillers have betrayed the rural communities they come from. 'I saw a local fella the other week and he was all excited, got a new ute, new skidder and excavator, the works,' Rousell says. 'He said he'd just got a contract from another contractor to pull out 1000 tonnes a week, and I'm like, why are you here then? Why aren't you working? And you know what he said? I've subcontracted the job out. The bloody sub-contractor had sub-sub-contracted the job out. What a dog.'

Coming across the paddocks from the main house is Megan, Rousell's wife. Wearing jeans and a T-shirt, she cuts a striking figure as she picks her way over the grass. Up close, she has perfect teeth and brown eyes. Like Terry, she also grew up in the area. 'I remember the myrtle forests we used to have, orange-gold leaves on the ground and moss growing up their trunks. People say Australia doesn't have seasons like in Europe, but they've never seen a myrtle forest in autumn.' She looks at the helicopter across the way, which has resumed its laps.

'When it all started, we wanted to leave as well. There are helicopters going over the whole time, logging trucks coming in and out, chainsaws whining. But where would we go? Who would buy our place anyway? Surrounded by this, we'd have to sell to plantations. And we love this place.'

Terry and Megan decided to fight. 'At Burns Creek, up the road, a friend of ours used to live on the creek, his kids swam in it during the summer. Then they got real crook so we went upstream to have a look and found all these dead wallabies in the water, frothing at the mouth.' They formed a community action group and similar stories emerged. One man was knocking in palings for a fence, aware of a buzz in the distance, when a helicopter was suddenly on top of him, spray-soaking him through. Another man, showering after he had been sprayed, accused his wife of buying new soap that was burning his skin.

The Rousells invited scientists and hydrologists to tour the area, but each visitor was ridiculed or ignored by the forestry industry and politicians alike – no matter what their credentials. 'One time a group of us were standing on Bodges Ridge and she was completely bald,' says Rousell. 'We had managed to get Dick Adams out there with us, a Labor MP, and he said, "I don't see any degradation." It was a steep ridge, completely cable logged. All the trees that held it together were gone and he was looking right at it – you just wanted to thump him. Then this smart man, a professor from the university, said, "Well, let's look at it from an economic perspective. Would this plot of land ever be of equal value as it was before it was cabled?" He couldn't answer that. They don't understand anything unless you talk in its economic relevance to them.'

But the community was worn down.

'After I got footage of a dodgy cable-logging operation, one of the bush bosses came to our front door and said if the footage ever left our lounge room, he'd send a few of the boys round to sort me out,' Terry recalls. 'I don't know what he was worried about. Seems everyone's got footage of that kind of stuff now and it doesn't make a difference.' But it was one of the last straws for him. By now he was driving petrol tankers for a living, and he worried too much about Megan and their daughter when he wasn't home. 'So we stopped.'

*

Terry takes me and his daughter in his four-wheel-drive beyond the Forestry Tasmania boom gates ('I made sure they cut me a key. This is public forest') and shows me the forest of his youth. He points out the odd stump with black rectangular indents where his father once knocked in 'shoes' – planks of wood for the loggers to stand on while cross-sawing a tree, one man on each side of the trunk and dragging a saw between them. The stumps are larger than a card table, and the black indents remind me of Ned Kelly's mask. 'Dad doesn't know why he bothered to do the job properly, not when they do this.' Every time we turn a corner, a gaping hole appears in the trees in front of us.

A couple of times, says Rousell, contractors have called him and confided that their crews had done a 'pigs' run' (checking out a coupe before starting work) through this forest and decided to 'accidentally' push over a tree with an eagle's nest in it to save going through the paperwork. When I ask why the contractor would bother telling him this, Rousell shrugs. 'Guilt, maybe. The guys on the ground are feeding their families and some of them might not like doing it, that's true.'

The car bumps along until Terry stops and gestures that I should get out. His little girl wriggles out behind me. I almost stand on a dead possum, and steer her away from the carcass as she lowers herself from the car. The smell of rotting flesh catches on the wind as we clamber up a steep slope. When we pass another dead possum, Rousell looks at it quizzically. I slip twice going up the incline. Each time, I steady myself by grabbing at the plants sticking up around me. Gorse. Thorns sting my hands and I pick them out of my palms, barely looking up until we get to the top.

I've started to get used to these clear-fells, but this one takes me aback. It seems to go on forever. Scraped bare, with prickly tufts of gorse sticking up like hair on a chemo-patient's head, the stadium-sized gully is pale and cracked. Beyond it, valley upon valley is much the same. There are thin lines of life where water must have once made its way slowly down the slopes. The odd fern is still standing, crisp-brown in the exposed sun. There's no way I could have seen this on my own. Even with Terry's car and his knowledge of the area, we still had to get out and walk up the steep track that only dozers and cable loggers can manage, to see the endless bald hills of the north-east.

'It doesn't matter how much rain you get now,' Terry says. 'Two days later it's gone. It doesn't stay on the land. The floor of the Tamar River is up three metres, and it's all our soil slushed down there.' The rivers come up too quick and they're gone just as quickly. Earlier in my travels I met a young journalist at the *Advocate*, the local paper for the north-west of the island. She told me about a time she and a photographer were sent out to do 'another tree-change story' – a feel-good piece, she explained, about mainlanders moving to Tasmania. 'So there was this woman in her backyard, dancing around with her arms up in the air, going, "It's so green here" – and we didn't have the heart to tell her all those trees behind her were plantations and she would have to look at massive clear-fells every fifteen years or so.'

In 2002, a Senate inquiry was set up to investigate the rollout of plantations in Tasmania. Embarrassingly, revelations within the inquiry suggested that the federal government's 2020 Vision had a rather slapdash approach to planning across the country, including no hydrological modelling to understand the water uptake of the proposed 200,000 hectares of plantations, no flammability tests and almost zero restrictions on what land could be converted to plantation.

For the Rousells, none of these revelations was a surprise. While the native forest behind them falls and boom gates go up, Megan has begun obsessively filtering their water. Their springs have dried up, Terry's sawmill is finished, and they can no longer swim in the river.

Back in their kitchen, I look at the forest outside. Terry makes me a sandwich for the road and Megan pours me a cordial. She looks out the window with me. 'There used to be devils and bandicoots and so many birds out there but they're gone. Since about '98 we stopped seeing them.' She hands me my drink. 'Now we've just the wrens, and they come for our garden, not the forest.' As I'm leaving, I notice a large tree on its side next to the abandoned mill. It's been cut strangely.

'Tilt your head a little,' Terry says with a grin, so I do. The log has been cut to look like a hand giving the finger. 'I'm getting an excavator to come in and cable it up,' he says. 'I want it to face the logging trucks as they come in and out of here.'

FORESTRY

'This debate has been resolved that many times over,' says Barry Chipman. 'They tell us to go into plantations, so we do, and now they don't want plantations. They want the timber industry to value-add, but they don't want a pulp mill. They are constantly shifting the goalposts.' Described by many as a moustache on a sapling, Chipman is skinny with white-brown gremlin tufts of hair and '70s-style beige spectacles. He has picked me up in Hobart at 7 a.m. and is taking me through the south-west, in particular the Styx Valley, which sits alongside the Florentine. A perennial defender of Tasmania's timber industry, he is a former timber worker – 'The de-barker. I used to take the bark off the trees with a crow-bar' – and is now the Tasmanian spokesman for Timber Communities Australia.

Renowned on both sides of the forest conflict for his folksy charm and bizarre analogies, he explains how the timber industry is a leader in renewable resources. 'Think of it like this,' says Chipman. 'You spill a carton of milk. Are you going to wipe it up with a sponge already full of milk? No. You're going to get a new sponge. It's the same with old trees. They're full of carbon. They're dying. Soon they will release their carbon. So we harvest it and store that carbon in products, while the new trees we plant suck in more carbon.' Giant trees are now carbon emitters. Old forests are *bad* for the environment. 'Out of all the primary industries – fossil fuels, oil, gas, you name it – we are the only industry that is renewable. We use solar energy and we *sequester* carbon.' By this logic, the timber industry is a kind of hero in the natural resources sector.

The UN's Intergovernmental Panel on Climate Change estimates that deforestation and forest degradation contribute 17 per cent of the world's greenhouse emissions and yet to date, Forestry doesn't have to include these emissions if the land is cleared and new, thirsty trees are then planted. It is true that furniture will store carbon for as long as the piece lasts. Paper, however, will start emitting its carbon in about three to five years. It is also true that new trees suck in pollution and release oxygen, but carbon is not only stored in the trunks of trees; it is also stored in the soil, the branches and the roots. Moreover, carbon cannot be separated from other functions of an eco-system, no matter how much the business world, with its attempts to 'offset emissions,' would like to do so. Long-established forests not only create rain but also act as filters, stopping the water from hitting the ground head on; without this filter, the rain will wash the top layer of soil downstream, which in turn clogs up the waterways and causes erosion. Scientists have recently discovered that the cool temperate native forests of Victoria and Tasmania, known as the moss forests, are potentially the most carbon-dense forests in the world. This has enraged the timber industry and Tasmanian politicians alike. When the Australian National University's *Green Carbon* report suggested that Tasmania's cool temperate forests should be left standing, the premier, David Bartlett, described it as 'bullshit.'

Hydrologists have also accused the timber industry of negligence, claiming it takes about 100 to 120 years for water retention to recover after a forest has been cleared. On a rotation harvest of 35 to 90 years, the water-table has no chance to recover. 'That's not true,' claims Chipman when I repeat this to him. He is driving me around the south-west forests of the Styx Valley and the Florentine. To prove his point, he takes me to a point where we can overlook a 'tapestry' of regrowth forests. 'They show you photos of a clear-fell but never this,' he says, waving his arm across the valley. 'It grows back, see? Sucking in carbon the whole time.'

The carbon argument is starting to get to me.

'Do you even understand this stuff?' I ask him. I confess that I have to look at diagrams and re-read definitions of the 'atmosphere, biosphere, hydrosphere and lithosphere' every time the word 'carbon' is mentioned, but it never seems to sink in.

He blinks at me when I stop talking and says finally, 'Well, no, not really.'

A voice pipes up behind us, 'It may as well be Italian to me.' Rex is standing next to the car. He is an older logger, brought out with us by Barry Chipman to tell me about the old days, but so far has barely got a word in edgewise.

I kick the ground with my boot, scuffing it.

'Same here.' There's a silence between us until Chipman's hearing aid emits a high-pitched scream and he wanders away to fix it.

*

In theory, Forestry Tasmania monitors logging in state forests to ensure sustainability. When Barry Chipman arranges for me to be taken on a tour of FT's headquarters, he says reassuringly, 'There are girls in there.' And he is right – there are women working there. There are also desks covered with lovingly cared-for plants, studies of beetles, landscape planners twitching their mouses, using digital modelling to map out forest plans, and a fire-management room where radio transmitters connect to watchtowers across the state. Written on the white board is 'Dogs ill from smoke, check with Vet.' I am loaded up with DVDs and booklets, filled with images of state employees out in the field – an interesting mix of older rugged men and young female 'earthy' types.

'There *are* good people working there,' geo-scientist Kevin Kiernan says of FT when I mention my suspicion that this wholesome image seems engineered. In the early 1970s, aged just twenty, Kiernan was one of the founding directors of the Wilderness Society. But as the organisation geared up for its campaign to save the Franklin River, he decided to study hard and work for Forestry instead. 'My theory is, don't join something, a group, you agree with. What good will you be there?'

Kiernan worked at the Forestry Commission (FT's forerunner) and then the Forest Practices Board as senior geomorphologist for fifteen years until resigning. 'In a sense it was a much more honest and direct place to work, to begin with at least. I encountered the reality, which was not that people in forestry were monsters and destructive, but they love their kids and want to look after them the best they can.'

As the commission's first geo-scientist, he worked on building up an overall picture of the island, mapping landforms, sand dunes and karst systems. But Kiernan found that science was one thing and decision-making another. He watched scientists present their research, only for absurd decisions to be made in Forestry's favour. For example, to prevent machinery from eroding the soil around streams, it was agreed that buffers should be put in place. But then the machines were allowed to reach over the buffers to pluck trees from beside the stream, thereby ripping up the root systems that held the bank together. Kiernan said he became more and more frustrated. 'They log with absolutely no concept of what is under-neath them,' he recalls. 'The landscape is what keeps everything – the trees, the roads – together, but they'll log the ridges, over limestone caves, karst systems, sinkholes. Tasmania is an inher-ently mobile landscape – you can't do things like that without even-tual collapse.' Tired of watching his research be disregarded, Kiernan resigned in 2002. 'FT was originally intellectually quite open,' he says, 'but gradually that openness was eroded.'

*

Last year Forestry Tasmania produced a commercial featuring 'Mark and Lisa,' a couple sitting in a marriage counsellor's office.

'Okay, Lisa,' the counsellor begins, 'you've read Mark's book. What did you think?'

'Well, it's a start,' she quips, 'but it could go a lot further.'

Mark throws up his hands.

'Oh gee, what a surprise,' he says drolly. Lisa's lips are turned downwards. She is one of *those* women – never happy, always nega-tive – in other words, a 'greenie.'

When I mention the ad to Chipman, he sympathises with Mark.

'They've got nothing nice to say,' he says of the industry's critics. 'They're never satisfied.'

Feeling like a marriage counsellor myself, I relay this to the Wil-derness Society's Gemma Tillack, an environmental scientist and campaigner. She furrows her brow. 'I'm just thinking,' she says after a silence, until finally, as if closing a filing cabinet in her head, she shakes her head. 'Nope. I don't think I've ever put out a positive media release about Forestry Tasmania in the four years I've been here – but I know the Wilderness Society definitely has.' After another silence, she says, 'I can see that Forestry has improved, but it has always been incremental – bite-sized pieces of change that we've had to fight for every bit of. The giant-tree policy – that's the result of a campaign. Protecting wedge-tail eagle nests – result of a campaign. Protecting anything at all? It's the result of a campaign.'

It does seem strange that the things the timber industry seems so proud of, at least in their advertising and on television shows such as *Going Bush*, are the things environmentalists have had to push hard for. Management strategies around streams, the cleaning of vehicles so weeds aren't spread, spotting eagle's nests, protecting the velvet worm – all these have come after outside scrutiny.

'And if you understand ecological systems and the extent of the changes the timber industry is making to these eco-systems,' Tillack continues, 'then you can see the industry has only come a fraction of the way.

'You can't help getting a little cynical about some of their changes and amendments,' she points out. 'They keep changing their definitions and every time the language changes, we have to change our asks. It makes us sound unreasonable and always upping the ante, but we mean the same thing.' For example, environmentalists can no longer refer to 'old-growth' without excluding a majority of the native forests. 'Now we have to say old-growth *and* high-conservation-value forests *and* diverse eco-systems. But they're the same thing,' she sighs.

Back in the white four-wheel-drive, I ask Chipman if the timber industry has gained anything at all from the environmentalists. In the backseat, Rex perks up.

'Oh yes, most definitely. Before they came along, forestry was disgraceful. Men were pushing debris into streams, cutting anything they wanted, roughing up the ground – the greens stopped all that. We do need them.'

Chipman clears his throat.

'Need*ed* them,' he says. Then, surrendering a little, he adds, 'I suppose they were necessary at the beginning, but not anymore.' Something switches in his head and he leans over conspiratorially. 'Like I said before, do you know how much the Wilderness Society raised last year?'

'Eleven million?' I say.

He nods, happy I've remembered. 'Earning that kind of money, I doubt they even want to save the forests.'

*

At the Big Tree Reserve, Chipman pauses at a tree around 300 years old and slaps the trunk like it's the flank of a racehorse.

'It's a little bit like walking through an old people's home, don't you think?' I laugh out loud in surprise, recalling the boiled-cabbage smell of my grandma's nursing home, the chain across her doorway to stop people with dementia wandering in and taking her things, an old man hunched over the phonebook desperately trying to find his name.

'Not really ...' I start to say, but he's moved on. 'Personally I prefer regrowth forest,' calling out over his shoulder. 'A working forest. Look how empty it is, there's no animals in here. Nothing for the animals to eat.' I decide against pointing out that this reserve is some 100 square metres in size and we can hear chainsaws and the thumps of trees falling in the distance.

How can so many people all be looking at the same thing and see it so differently? The man moseying around in front of me looks at a 300-year old tree and sees a nursing home, while an activist twenty minutes down the road sees a block of flats for furry and feathered creatures. Of the older activists I've met who fought to save Tasmania's Lake Pedder forty years ago, many speak of their belief that if people could only see something, take their shoes off and wade into it, they would feel the same way. But are these feelings – the instinct for natural beauty – universal? Is a sunset beautiful to everyone, or just some? And what of the modern sunset, made peculiar and at times spectacular by smog and pollution, is that also beautiful? And just how marred are our perceptions by our desire to be (or seem) something other than what we are? Back in Hobart, Chipman introduced me to a furniture maker, who took me to an exhibition. He showed me a Huon pine coffee table he had made, complete with rippled frosted glass. Patting it, he said, 'I am giving this tree a second life. I am storing carbon.'

'Um, it's really nice,' I said, before walking off to coo over a bench made from recycled wood and fishing ropes.

In the back of his ute, the furniture maker showed me some lengths of Huon pine he'd acquired.

'Smell it, isn't it beautiful?' he said. Flipping one over, I noticed it was carved in an Asian script. 'Oh, I make the "thank you" plaques for Forestry Tasmania to give to Ta Ann,' he explained. Ta Ann is a Malaysian company that partnered with FT in 2005 to set up veneer mills on the island and process state forest logs. I nodded, thinking *of course you do*. He was right about the perfume – the wood smelt gorgeous – but as I ran my hand over the scooped-out characters, I couldn't help thinking, why does it look so *tacky*? Isn't this meant to be 1000-year-old wood? Am I a snob, or is something dodgy seeping into this man's work?

'Our aesthetics are vastly different,' says Gemma Tillack. 'We're attached to the aesthetics of an unmanned forest, whereas the loggers love the manned forest, because it's their manning. It's their work and they've created it. It's probably the same feeling we get from our gardens.' Michael Field, a former Labor premier of Tasmania, wrote recently that the division in Tasmania over the environment flows from 'differing value systems, not some monopoly of the truth.' But is beauty a value? Is the environment a value – or is it, as environmentalists claim, a fact? Can our aesthetics differ to the bitter end? It's a 'brown leech-ridden ditch!' the Liberal leader Robin Gray spluttered in fury during the Franklin River blockade, unable to understand the passion environmentalists felt for the wild rushing river. Bob Brown recalls flying over the Franklin in a helicopter with the chief of the Hydro Electric Commission, who yelled over the noise, 'Look! It's not beautiful at all, is it?!' In a sense, this idea of natural beauty is the Greens' double-edged sword. It doesn't include *us* in it, which is part of its allure but also perhaps why we want to make a mark on it, to feel significant within it.

Pete Hay, a lecturer at the University of Tasmania and author of *Main Currents in Western Environmental Thought*, won't go as far as Tillack and equate what modern loggers feel for forests with love. Rather, he tells me over a coffee, they love how it makes them feel. 'At the negative end of the scale, they love it because it makes them feel like men; on a positive side, it makes them feel robust and able-bodied. But they love nature as a context for activity. Love the ocean because they can fish in it. Take away the fishing, and I assure you, they'll be less keen on it.'

But is there anything wrong with that? It's as if humans have become two-headed philosophical monsters, with one head sculpted by Descartes, who sought to break the connection between animal and man by dividing body from soul and reducing animals to the model of machines, and the other by Darwin, who sought to reconnect humans to animals by showing the inherent connection between all things. However, neither of these streams of thought seems to have induced in us a sense of responsibility. Instead, both have provided some handy excuses and justifications. Descartes affirms human superiority, while Darwinism seems to have produced a primitive selfishness, a primal *que sera sera*, whatever will be will be. Then there is a third head, the plain dumb one, that says, 'We just like to get in there, put on our headphones, play music real loud and smash shit up.'

*

The forest debate is a minefield. You need a bullshit detector to pick your way across it. Vitriolic and violent, it is as if some omniscient dramaturge feeds the conflict between these mere mortals – the loggers and the ratbags.

'Faeces,' Barry Chipman says emphatically about the activists at the Florentine blockade. 'They smear faeces on the workers' machines.' A week later, on local radio, he adds an ingredient. 'Faeces and *menstrual* blood on the machines.' And I understand now why I liked Tony Jaeger, the general manager at McKay's Timber, the man who drew on a sheet of butcher's paper for me. He didn't talk to me about the carbon-storage values of new-growth forest or unlocking the sunlight stored in a woodchip to generate electricity or pretend he cared about wedge-tailed eagles' nests. Nor did he divvy up the island into good and bad. Instead, he explained how to saw a Tasmanian log.

The evening after I meet with Jaeger, I find myself back at the Pink Palace with my head slumped on the dining table. I tell Ula, Wazza and some of the other activists that I'm not sure what I believe anymore. One activist tries to wear me down with rapid-fire statistics, but the core crew understands. As they cook a large pot of spaghetti sauce, they ask me to talk them through all the opposing arguments, and we spend the night pulling these arguments apart and seeing if we can put them back together. One activist suggests I should go up in a plane and see the island as a bird might. See the scraping scars, the burnt holes, tick off the ill-fated forests and the carpeted rectangles of monocultures. But, I say, I don't trust my own eyes anymore. Besides, how am I going to pay for a plane? I can't even afford to stay in a hostel down here. When Chipman insisted on picking me up at my 'hotel' at 7 a.m., I woke up at 6 a.m. at the Pink Palace and rushed to meet him out front of the hotel most journalists seemed to stay at.

Partially out of spite, I tell the activists that for all their independence and their determination to stay 'outside the system,' the ratbags may simply be pawns in a larger game. A couple stiffen, but Wazza nods.

'We try not to be used or baited when important forestry decisions or elections are going on. There's nothing the pollies like more than a little feral distraction.'

Loggers are also used to deflecting scrutiny from the top end of town. When the footage of Rod Howells allegedly smashing the car with Miranda and Nish inside made headlines, Bob Gordon of Forestry Tasmania accused the forest activists of taunting timber workers. But it is just as easy to read this metaphor in another way: these men have been purposely enraged by their employers and by politicians, and then turned loose on the activists. Why feed this rift? Why isn't this forty-year wound that divides 'greenies' and 'rednecks' allowed to heal? In whose interests is it to keep the island fractured?

I recalled a conversation at the blockade with Miranda. 'Tasmania has the population of a country town,' she said to me, 'but we have a government.'

I laughed when she said that, realising she was right. In my mind, I had kept adding an extra zero to the island's population because it seemed incredible that only 500,000 people live in the state. There are just twenty-five people, mostly men, in the lower house and fifteen people in the upper house, calling the shots. Some of them have been doing so for over thirty years.

'Maybe Victoria ought to govern Tasmania?' I joked.

'Or we could just admit it and let Gunns rule us,' she suggested. 'The Gunnerment.'

When I try to sleep on the floor of the Pink Palace that evening, with dogs nudging at the bedroom door and Ula coughing intermittently in her sleep, I realise I have to delve further. I go through the names of people I should speak to next – those off-stage meddling Greek gods of Gunns, for example. 'It's a bit like a monopoly board here,' an older activist had said to me. 'You go around and around until you have your utilities, where the gentry live and so on.' I'm only a quarter of a way around the board – and I'm yet to meet the island's powerbrokers. Who's in charge of this place? I've only just gotten comfortable with the loggers and the ratbags. Their lingo now rolls off my tongue with ease: quartersawn, pulp quota, direct action and black wallaby. Now I've got to wade further out, testing my footing and my ability to understand the machinations of this island. Politics. Governance. Economics. I curl into a ball and try to bury myself under the blankets.

I hear something thump out the front. Emerging from my hideout, I peer through the window. I can't be sure but it looks as though the Pink Palace's blue picket fence has finally snapped. *Nightmare tenants*, I think.

*

I've got the shits. Literally. And I'm driving diagonally across the state. Fuck fuck *fucking* ferals, I say, slamming my steering wheel with an open palm.

Feverish, vomiting and shitting, I have spent the past two days in Maydena. I came intending to speak to people but instead found myself sweating and paranoid on the main street. Clean your fucking dumpstered food. Fuck. FUCK. The man on his cement platform outside the petrol station seemed to watch my every move as I knocked on random front doors.

'I hate you I hate you I hate you' I chant silently to his little shop until I realise I am going to have to go in for supplies. Unsteadily I negotiate the steps and look at the faded packets of food. Tins of creamed corn, Barbeque Shapes and Coco Pops. Nothing. If this fucking island is meant to be the land of fine food, then where the fuck is it? Vending machine coffee, sour cream and chives chips or Monte Carlo biscuits. I buy two bottles of pretend juice and disappear into a motel room for two days. Dragging my body from bed to toilet, I lie on the bathroom tiles, enjoying a momentary peace after my mass internal exodus.

When I've recovered enough to drive, I head up the Midlands Highway. The sky is rumbling for a storm. The words 'I'm Back Robbo' are graffitied on a boulder next to the road, lingering like a threat to some poor bastard. Black swallows move like thumbprints across the sky and eerie totem poles gnawed by cockatoos stick up out of the scraggly paddocks. There are metal cut-outs of bushrangers, convicts and shepherds propped up on faraway hills and I put my foot on the accelerator, grinding the gears. I don't care if the island is logged to splinters. I want to lock myself to a hedge and scream 'Save Our Fucking Stupid Hedges.' I want to collect all the hi-vis vests in the world and bury them, dig them up and scream and burn them. Then collect all the polar-fleece Kathmandu vests and hiking boots and taunt, 'Bah Bah, I'm "Tasmanian," look at me, I can use a compass.' I'm sick and tired of bloody convicts. I want to yell at the next person who says 'I'm fourth-generation Tasmanian.' SO FUCKING WHAT? I'm 45th-generation Roman for all the fuck it matters. I want to grab the ferals and tell them to shove their anarchy up their a-holes.

When a wallaby strikes out across the road in front of me, I'm not prepared. Swinging the wheel, my car twirls into the other lane and then swings back. I screech to a stop in the emergency lane. In my rear-view mirror I glimpse the bum of the wallaby scampering into the bushes.

I drive on a little further. But when the sky cracks open, gushing out rain, I pull over. It's pitch black at 5 p.m. I scramble in the glove box for sleeping pills. I just want to shut my mind up for several hours. Taking three, I roll out my swag in the back of the car, blowing little puffs of cold air. Pushing away thoughts of serial killers and rapists, I think groggily, 'If anything happens, I'd rather be asleep.'

THE COMPANY

GUNNS

At first glance it seems the directors of Gunns don't care too much about the impression they're making in an environmentally conscious world. When I first saw their 'Little Green Book,' launched in 2009, I thought it was a prank. On the cover a butterfly, childishly outlined against a lime-green background, begins a journey through the various sectors of Gunns, from plantation to design centre, wood laboratory to nature reserve. One learns about everything from the protection of the rare Ptunarra brown butterfly to grape-growing. Each page is laid out like a picture book, with a spray-paint font and photographs digitally altered to just the right dawn or dusk light. As I flicked through the booklet, two conclusions suggested themselves: either Gunns was taking the piss, or some greenie was taking the piss out of Gunns. One thing seemed certain: neither party was taking the other seriously.

Nor am I taken seriously. My repeated attempts to speak to Gunns are not simply refused – they are ignored. When I call the phone number at the bottom of the company's press releases, I manage to get through to its spokesman, Matthew Horan, but soon discover he is not an employee of Gunns; rather, the company is a client of his. Based at a public-relations firm in Sydney, Horan agrees to speak to me at a later date but has since ignored my emails and phone calls. When I ask locals and state reporters if Gunns speaks to them and why its representatives won't speak to me, I get a reply that induces a sinking feeling: Oh, they will. After you publish.

But on a second, more in-depth look, I decide it's more likely that Gunns don't feel the need to do their own talking. Before he resigned as executive chairman in May 2010, one of John Gay's favourite responses in his increasingly rare media interviews was to attribute responsibility to the government body in charge of overseeing that chemical, that dam, that pipeline, and so on. Why won't they pay more for the state's trees? That's a question for Forestry Tasmania, he'd say. And why do you employ the forestry officers who inspect your coupes? Now that's a question for the Forest Practices Authority. And why did Premier Lennon recall parliament to push through legislation to approve your pulp mill? If that is what the government wants to do, then who are we to complain? In short, Gunns Limited has no real say in anything – it's just lucky, that's all.

*

The past twenty years have seen Gunns grow from a medium-sized sawmiller to a company that runs close to 85 per cent of Tasmania's timber operations.

Initially it was a little-known but steady venture started by two brothers in the latter half of the nineteenth century, its forest forays mostly unremarkable until the business started to focus on the state's woodchip operations in the late 1980s. Today the company leases vast swathes of state land, has first buyer's rights over much of the state's unprotected native forest and is the biggest private land-owner on the island. Gunns owns four woodchip mills, along with walnut farms, vineyards, veneer mills, sawmills and plantations.

It has owned a chain of hardware stores and sponsored events, sports teams, TV commercials and the prime-time weather forecast. It has its own stadium stand and owns construction firms; it even did the hundred-thousand-dollar renovation of Premier Lennon's home. On an island of 500,000 people, it's difficult to pinpoint where the company's influence begins and ends. And yet to see its red, black and yellow logo on the occasional sign while driving along the highway can come as a surprise, one gets so used to deciphering the company's invisible web of ownership and control. I once gave a bottle of wine to a Wilderness Society employee; she raised an eyebrow at me comically: 'You know that's a Gunns wine, don't you?'

It is often observed that the state has entrusted a great deal of itself to the timber company, but the directors of Gunns see things differently: they seem to have a pathological sense of being taken advantage of. Chairman John Gay repeatedly threatened to take the business to the mainland or even to China if locals placed too many environmental restrictions on the proposed pulp mill or asked for an increase in woodchip export royalties. His replacement, Greg L'Estrange, recently responded to questions about Gunns' sponsorship of a local cycling event with, 'We haven't finished our discussions, but certainly you would say our appetite for some of these areas has diminished. Life is a two-way street.'

Critics of Gunns say the company makes a vast profit from a publicly owned asset and gives little back to the island; its earnings, they argue, are siphoned back to investors and shareholders on the mainland, while the state cops the bill for massive water consumption, poor roads and internal division. In 2009 Graeme Wells, an economist at the University of Tasmania, prepared a report outlining subsidies paid to the Tasmanian timber industry. His analysis, funded by Environment Tasmania and the Wilderness Society, estimated that between 1997 and 2008 the commonwealth and state governments had provided the local timber industry with $632.8 million. In 2005 the Tasmanian Community Forest Agreement provided the local timber industry with $235 million dollars to compensate for a 'loss of resource' – some 148,000 hectares of state land. Conservationists later worked out that only 27,000 hectares of this area was actually threatened by logging. Wells' report conjures up a kind of banana republic where debts are socialised and profits privatised.

Gunns rejects this, claiming it brings millions of dollars into the Tasmanian economy. It pays tax, GST, land levies, stamp duty, licences, road tolls and wages. Getting a clear view of these figures, however, is difficult. According to the market research firm IBIS-World, in 2008 and 2009 Gunns paid $18.6 million and $14.9 million respectively in income tax. Gunns says that its long-awaited pulp mill will bring an extra $1 billion dollars in taxes, which will, images in Gunns' accompanying publicity materials suggest, flow through to the state's public hospitals and kindergartens. Such claims are impossible to prove, and, argue the company's critics, their impressiveness diminishes when one considers how much of this will be offset by the government subsidies the mill is to receive. In 2007, the Tasmanian Business Roundtable for Sustainable Development commissioned a team of economists to study claims made by Gunns and to assess the impact the mill would have on the state's economy. Their report found that the $834 million tax contribution over the life of the project – as outlined in Gunns' benefits analysis – would be cancelled out by government subsidies worth $847.3 million. Equally slippery are assertions that Gunns is the island's biggest employer, given that it insists on keeping its employee numbers confidential.

Since Gunns listed as a public company in 1986, journalists and forest activists have found that one of the few ways to get information from this opaque giant is to become a shareholder or gain proxy from a shareholder and turn up at annual general meetings. Unpractised at diplomacy, the company's directors have not handled this well; John Gay shut down AGMs on more than one occasion to avoid questions from the floor, only to have to re-open the meetings on the advice of company lawyers.

'Do you think you're on a blacklist as far as ethical investment goes?' Ticky Fullerton, an ABC journalist, asked Gay in early 2004.

'No, that doesn't really worry me,' he replied. 'From what I see about ethical investment companies, where they have investments in Australia they don't show very good returns.' In other words, Gunns does what companies are expected to do: it expands, turns a profit and pays taxes. Ethics doesn't come into it; it's not up to Gunns to rein itself in.

Whose job is it, then?

*

'Gunns is doing nothing wrong,' Naomi Edwards told ABC's *Four Corners* in 2004. 'If the government will allow it to buy its wood cheap, to trash the state, to sell it overseas, to move jobs overseas, then why wouldn't it? It's got an obligation to its shareholders to maximise returns ... The problem is not with Gunns. The problem is with Forestry Tasmania and the state government.'

An actuary, Naomi Edwards has written several independent reports on FT's finances. In 2000–2001 FT's annual report showed a 34 per cent increase in the quantity of state forest cleared – and an identical decrease in public dividends. Eighty-four per cent of all logs felled went to the chipper (the year before it was 94 per cent).

There were dodgy land swaps, and 120,000 hectares of public forest could not be found. When an employee at the Land Titles Office noticed huge and irregular chunks of crown land and state forest changing hands, and was instructed to withhold the titles from the valuation roll, he leaked copies of the titles to a local activist. A subsequent investigation by the *Australian Financial Review* revealed that FT had received crown land on the understanding that it would surrender land of equal value back to the crown; the Land Services Department, however, was unable to shed light on where this surrendered land lay.

But then in 2003, Forestry Tasmania reported its best year ever; they were exporting more pulp than ever before, bringing in a profit of $24 million. Edwards read between the lines. She found that FT had actually made an $11 million loss. The auditor-general came to the same conclusion, ruling that on fourteen out of nineteen performance indicators, it was the worst year ever for FT. According to Edwards, the $24 million profit figure was arrived at by ignoring standard accounting practices relating to 'self-generating and regenerating assets' (that is, assets like trees). FT's less publicised profit statement, released separately, clearly showed an $11 million loss.

Speaking to *Four Corners*, Edwards tried to explain what to her mind was doublespeak: 'What happened last year was that Forestry Tasmania got a lot of cash in the door because they chopped down a lot of forest. But they really damaged the value of the forest they left behind for future generations, and that's why their net profit was negative – because they were actually taking from tomorrow's profit [and] bringing it forward to today.'

Before moving to Tasmania, Edwards was a partner at the accounting firm Deloitte Touche Tohmatsu, and before that a director of Trowbridge Consulting, Australia's largest independent actuarial practice, until its merger with Deloitte in 2000. She has spent much of her career working on billion-dollar accounts, yet Forestry Tasmania responded to her criticisms by branding her a 'greenie' with no understanding of accounting or valuation. It was announced in parliament that Forestry Tasmania had lodged a complaint about her with the Institute of Actuaries. When Edwards followed this up, the institute told her they had never heard from Forestry Tasmania.

Despite having been accused of ignorance and manipulation, Edwards is more confused than angry. Why was a state-owned business allowed to run its assets into the ground? Theoretically the shareholders of Forestry Tasmania are Tasmanians. But what is good for General Motors is no longer good for America, and what is good for Gunns, judging by Forestry Tasmania's financials, is no longer good for Tasmania.

Under Labor in the 1990s, the seventy-year-old Forestry Commission began its transformation into a government enterprise. Of the island's 6.8 million hectares of land, the state forest agency was to manage 1.6 million. Its $272 million debt was cleared (or rather, the debt was now to be serviced by a consolidated fund, a.k.a. the public) and it was given a new name, Forestry Tasmania. Since 1998 native forest clearing has doubled in Tasmania and between 70 and 80 per cent of FT's harvested timber is sold to Gunns. However, while Gunns became one of Australia's top-earning public companies during this period, dividends to the Tasmanian public halved. In 2007 a revaluation of Forestry Tasmania's 'biological assets' revealed a loss of $38.5 million for the year despite (or perhaps because of) an increase in logging.

Forestry Tasmania appears to be recklessly running through

its only assets for short-term gain while pretending to be something else entirely – a community-driven, tourism-focused, sustainable manager of forests. FT likes to blame its losses on its community-service obligations. There is the upkeep of walking tracks, picnic areas, tourist facilities and signage, as well as studies of the velvet worm and of new methods of clear-felling; there are books, DVDs, posters, commercials and a TV show to produce. But if FT has obligations to the community, why is it propping up a largely unskilled and diminishing timber workforce to provide woodchips for Gunns?

Providing approximately half of Gunns' timber, one might expect the state forest agency to have some power in dictating the price it receives – but with one tonne of woodchips from the state-owned forests currently going for $12 to $14 a pop (which Gunns then on-sells for an average of $150 per tonne), it seems that FT is only just managing to cover its costs. As timber craftsman Kevin Perkins suggests, a single piece of high-end Tasmanian timber furniture could sell for the price FT currently receives for 1400 tonnes of woodchips – but the state forests are appraised according to the market prices of a low-value, high-volume commodity.

Naomi Edwards describes the commercial relationship between Forestry Tasmania and Gunns as a 'monopsony' in which the latter has the ability to dictate price to its supplier. FT has no real alternative buyer. In the past decade the state forestry agency has attempted to establish commercial independence through its venture with Ta Ann but remains largely at Gunns' service. It is a peculiar thing. Why *pretend* to have a good business when you could actually have one? It doesn't make sense. There must be an underlying logic to all these transactions. But what is it?

In fewer than seven years the island's Regional Forest Agreement will expire. The 1997 agreement and its subsequent amendments were meant to induce Tasmania to restructure its timber industry. However, two years ago Forestry Tasmania secretly signed another twenty-year wood supply agreement with Gunns. The agreement will not only double woodchipping, but also promises Gunns $15 million in compensation if any future legislation interferes with this. The agreement sets a 'floor price' of $12.50 per tonne of native woodchips, but there is to be no 'ceiling price,' and Forestry Tasmania has indicated that it expects to enjoy a 'substantial bonus' if world pulp prices spike. According to Naomi Edwards, however, pulp is a highly volatile commodity. She says that pinning the value of the state's native forest to the global market price of pulp is even more risky than the current arrangement, whereby it fluctuates according to global woodchip prices. At the proposed pulp mill, 4 tonnes of woodchips would produce a tonne of pulp. The current market price of bleached hardwood chemical pulp is just over US$750 per tonne. In mid 2009, it plunged to US$480; in 2002, it dipped to US$200 before shooting up to US$500. 'To survive the global roller-coaster export pulp cycle,' Edwards wrote in her submission to the Resource Planning and Development Commission when it was assessing the proposed mill, 'Gunns will have to aggressively manage its labour costs (including harvesting contractors) and its wood costs.'

In 2004, after watching both Labor and Liberal wax lyrical about FT's balance sheet, Edwards realised there wasn't going to be much truth-telling on the floor of parliament. 'I think a Royal Commission really needs to ask why would a government let this terrible situation happen to its own state, with destruction of the environment, of financial returns, and of jobs.'

*

Gunns has a wide range of supporters. There's the tenacious Barry Chipman of Timber Communities Australia (TCA), who seems to have every politician, reporter and media boss on his speed-dial, defending the industry like a chihuahua on pseudoephedrine. There is FIAT, the Forest Industries Association of Tasmania; John Gay was once chairman, and later joined its board representing Gunns, its largest member. There is FIAT's peak body, NAFI, the National Association of Forest Industries, of which Gunns is a core member. There's the forestry arm of the CFMEU. And when the company really wants to call the dogs in, it looks to the timber contractors.

All of these self-proclaimed independent bodies are linked by one thing, directly or indirectly: Gunns' money. Many aggressive political advertising campaigns can be traced back to FIAT. In Bob Burton's *Inside Spin*, an award-winning investigation into Australia's public-relations underbelly, Burton discovered that of the TCA's total revenues of $836,977 in 2002–03, $723,154 came from direct industry contributions. In the same year, Barry Chipman's wages were paid by NAFI. Even some of Gunns' political donations are paid in increments, appearing to shadow government decision-making.

There is a catch for spokespeople like Barry Chipman and organisations like FIAT, one that snags repeatedly on their reputation. 'Totally astroturf,' one activist says to me when I ask about Timber Communities Australia. Yes, Chipman is real and he can get a mini-busload of rural folk to a rally, but TCA is regularly accused of being an industry front masquerading as a grassroots community group. Katherine Wilson, in her *Overland* essay 'Grassroots Versus Astroturf,' defines astro-turfing as 'the creation of bogus community groups or independent authorities who endorse industry practice, recruit lesser-informed citizens, confuse the debate and make the real community groups appear extreme' – until it's difficult to tell who's who and what's genuine.

In 2003, for example, the Tasmanian branch of TCA was accused of hijacking the Preolenna Mothers' Group in the island's north-west. A local group had organised a rally of a few hundred people to protest in Burnie against the proliferation of plantations. They followed this up with a submission to the federal Senate inquiry into Tasmania's plantations. However, by the time the investigating senators organised to visit the township of Preolenna, it seems Chipman had got there first. The Preolenna Mothers' Group was now the 'Preolenna Mothers' Group, Timber Communities Australia Pty Ltd.' The local hall's lease was even in TCA's name. The senators were fed scones and tea and told that the community of Preolenna, albeit smaller, was perfectly happy.

The union movement has also come under scrutiny for its relationship with Gunns. Neal Funnell, an activist and lawyer in his mid-twenties, suspects an unhealthy relationship has developed between the forestry unions and their members' largest employer. In 2007 he released a paper alleging that industry had taken 'steps … to ingratiate, financially indebt and then control the forestry union.' Funnell interviewed union members from the CFMEU's national executive, and was leaked several documents from its top ranks. He found that the forestry branch had clandestinely been taking money from industry, via the National Association of Forest Industries (NAFI), to offset mounting legal costs. In 1999, loggers had surrounded a campsite occupied by conservationists in Victoria's Otway Ranges, essentially holding them hostage for four days. Under physical threat, nobody was allowed to enter or leave the camp unless they signed a homemade 'contract' promising they wouldn't protest in the area again. Afterwards, activists launched an assault case against the loggers. The CFMEU was the first defendant and found itself in a million-dollar lawsuit. In 2005, according to Funnell, one union official was said to have bragged that Gunns had given him $600,000 to bail the union out.

Around the same time, an organisation called Earthworker, established to try to mediate between unions and environmentalists, had been making headway. Less than a year after its inception, fourteen unions and three environmental groups had affiliated themselves with the new organisation. It aimed to bring green groups and union members together around the goals of solidarity and sustainability. The new movement was about to go national, but the Otways assault case proved too much and the fragile coalition dissolved. 'The timing of Gunns' $600-k pay-off [to the union] on this occasion was no accident,' Funnell has claimed. 'The largest logging company in Australia had just indebted the Forestry Division of the CFMEU in a federal election year where forest policy was shaping up to be a pivotal issue.' The status quo was restored, with the unions back where the industry wanted them to be.

Then there is the matter of Gunn's influence in state parliament. During his short-lived stint as leader of the Liberal opposition in 2001, Bob Cheek recalls how he once naively chatted to a journalist about what he considered the state's impotence when it came to forestry, wondering if there was 'room to move' on the current policy. The following morning his 'pro-green' stance was on the front page of the newspaper and Cheek came to work to find a sullen circle of furious ministers. His greatest internal opposition came from the shadow forestry minister, Rene Hidding, who not only relied heavily on pro-timber votes in his electorate, but had also, before entering parliament, sold his own businesses – Hidding Trading, Hidding's Mitre 10, Hidding's Building Services,

Span Truss Systems and Hidding's Joinery – to Gunns and had been employed by the timber giant to enable the transition. Another of the furious ministers was married to a plantation owner, while yet another held forestry contracts.

Then Cheek received a phone call from John Gay. 'He wanted to see me in Launceston that night – and said it "would be worth my while,"' Cheek recalled in his memoir, *Cheeky*. 'By the time I got to Launceston it was late at night and Gunns' headquarters were deserted – except for the CEO in his modest office.' That evening, Cheek says, John Gay dangled two prizes before him – a guaranteed cheque for $10,000, plus another $20,000 to come 'if I locked in the right answer to the question: "Will you continue to support the existing forestry policy?"' The Liberal leader says he took the $10,000 with no strings attached. He refused the extra $20,000, but told Gay most of his party was against his policy to end clear-felling, so he would probably support the status quo going into the election. When this story was put to Gay, his office said that Gunns complies with all laws relating to political donations, and that Gay had a very different recollection of the meeting.

As it happens, the Liberals' support of forestry hasn't really gotten them anywhere – their Labor opponents are that little bit closer to the main players. No doubt Labor's claim to represent the 'working man' is a useful perk for industry when it comes to bolstering union support against a 'common enemy' – the greenies. But Cheek's experience reflects what many see as Gunns' unhealthy influence on both side of politics – an influence so pervasive, some cynics claim, it is more useful to talk not of Labor or Liberal, but of the Gunns Party.

THE GUNNS 20

In 2008 the Still Wild Still Threatened crew visited Gunns Limited. They entered the company's head office in Launceston carrying bags of woodchips they had scavenged from university garden beds and scattered them in the foyer. They were occupying the premises, Ula announced, as she speed-dialled local reporters. Managers emerged from their offices and gathered around her, demanding that she name the endangered species she was so worried about. She listed some of the trees she believes are endangered by commercial logging. Mid-sentence, one manager interrupted her and asked for the plants' botanical titles – 'Anyone could know their common names,' he sniffed. When a noise on the roof startled them, one employee jogged outside just in time to see a bed-sheet unfurl across the roof tiles. 'The Real Parliament,' it declared. In the car park below, the media had arrived with their cameras.

'The hero of the day was their secretary,' Ula recalls. 'She kept answering the phone, putting people through or on hold without batting an eyelid.'

When the police showed up and escorted the protesters out of the building, they waved goodbye to the glowering employees, leaving the bed-sheet draped on their roof.

'They couldn't believe it,' says Ula. 'People have been too scared to face them directly since the Gunns 20.'

In 2004, Gunns sent out twenty writs to seventeen people, two environmental organisations and one community group, most arriving on Christmas Eve. The company was claiming that the combined effects of nine protest actions had cost it millions of dollars. Now it wanted it back, to the tune of $6.3 million. Bizarrely, Gunns was accusing the defendants of 'conspiracy' – although many of them had not met until brought together as members of the 'Gunns 20.' They were an eclectic bunch: a doctor who had just been named supervisor of the year at the Royal Hobart Hospital (he had raised public-health concerns about possible bacteria growth in Gunns' woodchip piles), a dentist dedicated to protecting the Tarkine, two filmmakers, Greens politicians, a musician, a law student, local residents rallying against clear-felling in their sleepy valley, activists who'd stopped work at woodchip mills, and organisations that had lobbied Gunns' shareholders, investors and customers. All were now in the same legal boat.

It took a matter of weeks for a counter-campaign to be launched, with defendants dubbing themselves the Gunns 20. 'So Sue Me' stickers were printed and stuck on cars, and defendants used them to tape their mouths shut. Critics called it a classic SLAPP suit, short for Strategic Litigation Against Public Participation, an abbreviation three of the defendants told me they could not say for fear it be added to the list of their indiscretions. Activists see this type of litigation as a David and Goliath battle fought on Goliath's home turf, with Machiavellian lawyers paid top dollar to squash dissenting opinion. The company would counter that it has the right to defend its reputation.

To begin with, the 216-page writ was dismissed by Justice Bernard Bongiorno as legally 'embarrassing' and 'ambiguous at best and misleading at worst' – but Gunns kept at it. Then, after two years of legal wrangling and three changes of lawyers, the company put forward a revised writ for the fourth time. It held. Five of the defendants had been dropped, including Senator Bob Brown and the Greens politician Peg Putt, and Gunns was ordered to pay their costs. Allegations of conspiracy and damages against the fifteen remaining defendants were finally allowed to make their way through the courts. A lawyer for the defence, Julian Burnside, said of the company's persistence, 'It leaves you wondering if the purpose is simply to terrorise.'

SLAPP suit or not, the litigation was undoubtedly strategically timed. Just days after issuing the twenty writs, Gunns released its proposal for a $2.5 billion pulp mill. The legal wrangling occupied the Wilderness Society during a crucial point in forestry negotiations. And whether it was the company's intention to 'terrorise' or not, its relentless pursuit and near limitless resources did intimidate its defendants and other potential critics in the wider community. Even as various defendants were dropped from the writ over the next four years, the timber company succeeded in dragging most of them through six expensive and exhausting years, all the while costing Gunns at least $3 million in court fees, legal costs and out-of-court settlements. Which is why, when Ula texts me in early 2009 to tell me she is being sued by Gunns, I worry about the hint of pride I detect. 'We're the Triabunna 13.'

Ula is not being sued for her 2008 visit to Gunns' headquarters, but for a mill action. Work was stopped for seven hours at the Triabunna woodchip mill when three activists locked on to the conveyer belt and four to machinery; the company estimates the stoppage cost them half a million dollars. Images of activists, tiny as ants, attached to enormous machinery against a deserted background of woodchips at dawn reveal the steeling of the young activists' faces as they readied themselves for the abuse. A line of revving log trucks ran for a couple of kilometres, their occupants baying for blood over their two-way radios as they waited to unload their trays of logs.

'Sluts, sluuurrrts, slurrrts, I wouldn't dessicate on you,' the spokesman for the Triabunna woodchip mill hissed in Ula's ear as she and two other female activists prepared to speak to the straining media circle of cameras and dictaphones. 'I think he meant "defecate,"' Ula says, raising an eyebrow. Gunns later asked the Tasmanian police for the names and addresses of all the activists arrested that day, no doubt hoping to avoid the embarrassing stuff-ups of Gunns 20, when some defendants were accused of attending actions when they were not even in the state. The police, breaching confidentiality laws, complied.

Later I am shown a photo of the Triabunna 13. Some are dressed in suits too big for them, which billow like cheap black and brown sails. They stand in a line with their legs slightly apart, like footballers. They've tough and sombre faces, as if they're thinking about the sleepless nights that lie ahead of them. It's the look one gets when one knows something bad has happened but can't help enjoying the attention. A lawsuit like this is recognition: the protesters have been noticed, like a mosquito being swiped at by its target. Among some activists, I suspect, being sued by Gunns is a badge of honour.

*

Heidi Douglas, one of the Gunns 20 defendants, nods when I suggest this to her. 'Yeah, there were a few people who felt they deserved to be one of the twenty defendants more than me. Some were actually angry they hadn't been singled out.' She sympathises. 'Definitely there were people more "worthy" than myself,' she laughs.

We are meeting in a Hobart café. Heidi, thirty-one, is wearing an oversized striped woollen jumper, her small white fingers knitted around a cup of tea. She says there was antagonism from some of the Gunns 20 when she decided to settle and sign an agreement with the company.

'Some saw the lawsuit as part of a broader campaign, which is fine, but I couldn't handle it.' She was just twenty-five when she received her writ, and Gunns was claiming $464,313 in damages against her. A filmmaker working part-time for the Wilderness Society, she shot videos of two protests in the Styx Valley and at a woodchip mill in the island's north-west. She left the mill when asked, but was later charged with trespass and found guilty. 'Gunns said we were all in this big conspiracy together, but we were, are, all so different.'

Heidi says she became depressed. 'I'd been planning on buying a flat. I really wanted a home, to be still, and then the writ came through and made me rethink everything.'

Rather than a badge of honour, for Heidi the lawsuit was an exhausting exercise. She felt like her life had been frozen in time, and was haunted by the very real possibility of everything being taken away. After four years of negotiations, Heidi paid no damages but signed an agreement not to lock on to machinery, halt operations on Gunns' land or trespass on its woodchip mills until 2012. She doesn't seem to mind.

'I think for some of us the case affirmed we need to take sides and dig in, but for me, I think I've learnt that this, all of this, needs to be healed. My films were one-sided and I'm beginning to see things differently – not my core beliefs – but how to make a steady and coherent analysis.'

*

In 2009 Gunns settled its high-profile writs against the Wilderness Society, its director and head campaigners, with both parties claiming victory. The Wilderness Society was ordered to pay Gunns a total of $25,000 in damages, as well as $15,000 from its former campaigner Geoff Law, while Gunns was to pay the organisation's legal costs to the tune of $350,000. Wealth is, however, relative. The case is said to have had a chilling effect on the Wilderness Society. In its aftermath, an internal rift has opened up between campaigners and employers, in part over the Wilderness Society's newfound shyness about protesting. In time, ten other defendants' writs were also settled, their costs paid by the company, with some activists agreeing to sign guarantees like the one Heidi signed, and others issuing carefully worded public apologies.

One of the youngest defendants in the Gunns 20 saga was Ben Morrow, a close friend of the activists at the Florentine blockade. In the months after receiving the writ, he was diagnosed with bowel cancer. He was twenty-eight at the time. 'I've seen all of us burst into tears in anger and stress,' Heidi recalls. 'But Ben, he was dealing with so much more.' When Morrow contacted Gunns to explain his situation and ask if it would drop the case, at least until he recovered, the company refused. In Melbourne for a court date, he collapsed and was rushed to hospital. Doctors immediately hooked him up to tubes. His former employer, Greenpeace, stepped in and took over his defence so he could rest. Late last year he died. Many of the Gunns 20 attended his funeral.

As the fifth year of court proceedings and negotiations wore on, more defendants were released from the case. Neal Funnell, the author of the paper alleging unhealthy links between the forestry industry and the forestry union, was one of the last to have his case dropped. It is no wonder Gunns had Funnell in their sights: the young activist was already infamous for a stunt he pulled off during the launch of a new ferry between Sydney and Tasmania in 2004. He and three other protesters managed to foil the ferry's maiden voyage as it headed out of the Sydney Harbour by abseiling down the side of the boat and neatly hanging a banner alongside the ferry's name, transforming it into *Woodchipping the Spirit of Tasmania*. The media loved it. Twenty-three at the time the writs were served, he was the youngest defendant. By the time he was called on to appear in the Supreme Court in 2010, he was an experienced lawyer, and announced that he would be representing himself. Gunns dropped the case less than a week later.

And then there were four.

The remaining defendants could not make the same show of strength as the original Gunns 20. Exhausted, with few assets, they were anxious about taking more time off work and one of them was seriously ill. They were all residents of Lucaston, a cosy valley forty-five minutes south of Hobart. Of all the Gunns 20 defendants, they seemed to have the most direct reason to oppose Gunns. Their community had been notified that 1000 hectares of privately owned forest near their town was to be logged.

'I feel like a fraud. I'm not an activist. If anything, I'm a bit of a nimby. Not in my backyard you don't,' Lou Geraghty says when I drive down to Huonville to meet her. A tiny woman with puckered lips, I find her in the café kitchen of the local environment centre, burning several veggie burgers.

'I like to build kitchens, not cook in them,' she explains, 'but I'm taking work wherever I can get it.' Geraghty, in her late fifties, is a grandma and not usually the protesting type. In fact she had been close to the neighbours who were selling their land to Gunns.

'Jimmy and I were good friends,' she says sadly. 'People, his kids, would always come to our place to find him. He feels like I betrayed him, that I led this whole outcry against him, but I didn't. He was the one leaving us to watch the entire place get logged, burnt, baited and sprayed. And then live in it.'

Geraghty grew up in Eden on the coast of New South Wales, the first Australian port to export native woodchips. Japan even named a cargo ship, specially built for the woodchip trade, in honour of the town – the *Empress of Eden*.

'There wasn't any attempt to hide it back then,' Geraghty recalls. From their house overlooking the valley, her family watched as the tops of hills, sloping valleys and wet gullies were shaved down to the soil. 'The whole place was razed for woodchips.' When she moved to Lucaston almost thirty years ago, after falling in love with a cheap block of land bearing ferns, streams, mossy rocks, tall eucalypts and waddling wombats, she never dreamt it was going to happen all over again.

Despite what they were told, most residents knew life would change in Lucaston if the surrounding forest became a logging coupe. There would be poison baits and shooting parties, post-logging burns, aerial spraying, a constant rumble of log trucks and declining property values. The views outside their windows would be permanently altered. A group of locals – the future Gunns defendants – roused their neighbours and began to petition against the sale. They wrote letters, launched a website and held meetings with politicians and with Gunns. They even managed to raise the same amount of money Jimmy was getting from Gunns but, furious, he refused to take it. Finally, when each rational approach had failed, the residents set up a blockade. Then one evening Lou Geraghty's phone rang. It was John Gay of Gunns.

'I got the fright of my life,' she recalls. 'I was so shocked, I can barely remember what he said. He was just yelling.' A few months later, her name and those of two other locals appeared on the writ. The fourth name was the Huon Valley Environment Centre itself.

'I guess I should've known then something was up,' she says of that strange phone call. 'That was John Gay's way of warning me.' This rabble of residents had managed to save Lucaston from the skidders, only to face losing their homes in the court case. When I ask her what is at stake for her, Geraghty says sadly, 'My home and my bush block. And they'll log it first chance they get.'

But then on 29 January 2010, Gunns announced it was dropping its case against the remaining defendants and would pay them $155,088.

Despite the seeming futility of it all, the company is continuing proceedings against the Triabunna 13, demanding damages for the disruption to business and also for actions designed to 'gain media publicity.' Still Wild Still Threatened have responded with a counterclaim, alleging the timber company and their spokesman have misled the public with claims that no old-growth forest will be used in its proposed pulp mill, when harvesting plans show the opposite. 'Gunns have consistently claimed that their proposed Tamar Valley pulp mill will not use old-growth logs. Our counterclaim states that Gunns have used misleading and deceptive representations, which constitutes a breach of Section 52 of the Trade Practices Act,' Ula announced to reporters.

*

In 2005, the wife of one of the Gunns 20, Peter Pullinger, stood up at the Gunns annual general meeting and asked the directors of the board to withdraw the $784,000 claim against her husband.

She held up a photo of her family and described the toll the writ was taking. Locals for thirty years – Peter Pullinger as a dentist in Burnie – the Pullinger family had campaigned to protect the Tarkine, a campaign that was largely won. The lawsuit against Dr Pullinger alleged he had attended and organised meetings and raised concerns about the health risks of the woodchips on the Burnie wharf. When she went to the AGM, Leonie Pullinger wanted the Gunns directors and shareholders to see that she and her family were not 'ratbags.' Phillip Pullinger, her then 24-year-old son, a practising doctor who was later awarded Young Tasmanian of the Year as founding director of Environment Tasmania, was waiting for his mother in the foyer outside the meeting when two men approached him. Open-palmed, they shoved him into the far corner. 'You're dead,' they hissed, standing over him until the doors opened and people were flushed out of the meeting.

Afterwards, John Gay told the press he was sorry Mrs Pullinger had been affected by the lawsuit. 'I have the same feeling for her as I have for my family and my employees,' he said. 'The green movement have done enormous damage to families and employees of this company ... My own family have had phone calls annoying them and saying nasty things to them, and letters. I'm very sorry that she is in there, but they should have thought about what they did before they did it.'

Despite the proliferation of sassy 'So Sue Me' stickers, Gunns' lawsuits have made others think twice before criticising the timber company. Protests against the proposed pulp mill have been run past lawyers and re-framed so as not to target Gunns directly; instead, they focus on the state government's involvement in the approval process for the mill. In the Blue Tiers, in the state's north-east, a key environmental organiser received numerous phone calls from her neighbours, begging her to shred pamphlets and meeting notes and to close any bank accounts linked to their environmental work.

'I kept assuring them that we hadn't done anything wrong,' she told me, 'but people were like, "That doesn't matter, the Gunns 20 didn't do anything wrong either."' With almost half of the island's wine industry at risk if the pulp mill goes ahead, the wine-buyers at the Victorian Sommeliers Association considered boycotting Gunns' wines but decided against it. The decision, their president admitted, was directly influenced by the Gunns 20 lawsuits.

Employees of Gunns have also suffered, claims John Gay. A few weeks before Christmas 2009, he and his wife were woken in the early morning by a ruckus in their affluent Launceston street. Fences were being slammed, intercom buttons pressed and things pushed over. 'Greenies,' they immediately thought. It was not entirely surprising that the Gay family would be targeted: it's said they've been spat on, yelled at from passing cars and even refused service in a restaurant. By mid-morning, police were looking for fingerprints and inspecting a small burnt hole in the doormat. The local media reported that Gay and his family had been besieged in their own home; the *Sunday Examiner* claimed anti-pulp mill campaigners had attacked the house with a 'smoke bomb' and gave the former premier, Paul Lennon, a whole page to vent his outrage. The incident, he wrote, was yet another 'terrorist activity by pro-testers who think they are above the law ... Vigilantism has been part of the tactics of anti-pulp mill activists for some time.' For-estry Tasmania released a statement describing the vandalism as an enormous setback for the anti-forestry movement.

Two days later, police revealed the incident was a drunken prank by several young men who had spray-painted dicks and balls on fences along the street. A twenty-year-old man was charged; he had no connection to the anti-pulp mill campaign. Only the *Mercury* reported these developments. By then, the damage was done and the island's media refused to undo it. Buck Emberg, a vocal member of Tasmanians Against the Pulp Mill, couldn't help noting a double standard. He has found a deer's head in his driveway and dead fish in his letterbox, and has been forced off the road three times by logging trucks. None of these incidents was investigated by the police. But when someone sprays a dick and balls on John Gay's fence, they send out the forensics crew.

MUSICAL CHAIRS

When challenged about the company's environmental record, Gunns' spokespeople invariably deflect the blame, pointing out that the company is simply taking commercial opportunities where it finds them. As John Gay told the *Age* in 2003, 'If I rejected [the chance] to take some logs, they [Forestry Tasmania] would just issue them to someone else.' In other words, the company was simply operating within state and federal law. '[The protesters] can keep coming, but we don't make the decisions.'

Is that what it comes down to – is the government ultimately responsible for everything that happens in Tasmania's forests? The state forests are a public resource, controlled by a government-owned company. But when it comes to Gunns and Forestry Tasmania, it's impossible not to wonder: is the dog wagging its tail, or the tail the dog?

'How can you avoid that?' says Paul Lennon, Labor premier from 2004 to 2008, when I ask him about the closeness between the state government and Gunns. 'In a small community such as ours, you're bound to have one or two companies dominating – we can't just ignore them because people think it's unethical for us to meet … We're a small government – we have to make decisions and take responsibility for market and industry that other governments can leave for competition to battle out. There does need to be amendments, such as the open-ended cheque, but we don't have the luxury to not be involved.'

Known as Big Red, Lennon is a large, ginger-haired man with jowls like saggy red steaks. A former union boss, he was known for his ability to appear bored and half-asleep during question time, only to launch into a snarling attack, rage rising like a rash up his neck, at the slightest nudge.

It has been suggested to me that his temper is a bit of a ruse. Pete Hay, a former Labor adviser, describes the 'concocted fury' of politics as a sort of tool of diplomacy. 'I had it too,' Hay admits. 'I remember one time I tore a man to shreds on the phone with a huge grin on my face while others listened in. You'd feel untouchable all week and then on the weekend, you'll review what you've done and feel terrible.'

Now, as I watch Lennon walk into the coffee shop where we're meeting, I feel a twinge of sympathy for him. He strikes me as a man who has put all his power into his bite, his brutishness, only for it to diminish with age, leaving him a toothless tiger. His rage has left a trail of rattled people in his wake; stories from his union days have him throwing punches at rallies, heckling speakers, starting spats at the urinal; during his ministerial career, he's said to have thrown people against walls during meetings, and to have drunkenly abused an environmental campaigner in a Launceston restaurant; the unlucky activist was out to celebrate his fortieth birthday.

Lennon, I discover, has an uncanny ability to slam the door on a line of questioning with a simple, stubborn, 'No, I didn't.' Only when you are back at your desk, looking over your research, do you say exasperatedly, 'But you did!' In a profile of Big Red in *Good Weekend*, Richard Guilliatt described one such conversation:

'Aren't Tasmanians,' I ask, 'entitled to be wary, given that there's a history of corruption involving politicians and the forestry issue?'

'Corruption where?' Lennon shoots back.

'Well, you had a Royal Commission here 10 or 12 years ago ...'

'What's that got to do with forestry?' he challenges.

Well, Edmund Rouse was the chairman of Gunns.

'Chairman of Gunns?' Lennon's eyes narrow. 'Are you sure?'

Erm, that was my understanding.

'Aw, give us a break, mate.' He throws up his hands. 'I think we've just about had enough, haven't we? Give us a break!"

Now, as I question him, Lennon ogles me. It's unnerving. I ask about the 2004 federal election.

'Why did you appear to support Howard, the Liberals, over Labor?'

His moustache wriggles a little.

'No, I didn't.'

And it works. I'm no Kerry O'Brien. Instead I stammer, 'Really? I thought you did at the beginning before being told ...'

'Nope,' Lennon shakes his head. 'Definitely *not*.'

Our conversation limps along. I figure Lennon has agreed to talk to me because he feels he has been portrayed poorly in the past, but he's not helping. It is as if he only knows two types of interactions with a journalist: talk at length about some unquotable bureaucratic process until the time is up, or react, defend, attack, react, defend and so on. But I want to hear about the times he and the Greens *have* agreed – haven't they ever got along? He narrows his eyes, as if this is some type of reverse psychology, which it might well be.

'People have as much in common as they don't,' he responds cryptically.

Could you give me an example? I ask.

'Putt and I agreed on some things,' he says, referring to the former state leader of the Greens.

Such as?

'Social issues.'

Okay.

He looks bored. I suspect I've disappointed him. I disappoint myself in situations like this. Why can't I just attack? Play political chess? Wind him up? But then I realise, as he fiddles with the sugar sachet and half-mumbles under his ginger moustache, that he has disappointed me as well. I was expecting the toothless tiger – that was no surprise – and I had no doubt he would be smarter than he is usually portrayed. What disappoints me is the realisation that our conversation is no different from any other interview he has given. That he probably knows no other way to communicate.

I ask about his famous temper.

'My biggest weakness was loyalty rather than temper,' he says, twisting the question as only a politician knows how. 'I backed people when I probably should have let them loose. Bryan Green and Steve Kons, for example. That would be my biggest weakness, backing people to the hilt out of loyalty.'

In his four years as premier, after taking over from Jim Bacon in 2004, Lennon lost two consecutive deputies to scandal. Lennon's second deputy, Bryan Green, who took over from Llewellyn, was forced to step down within a few months; he was accused of brokering a backroom deal with ex-Labor ministers, enabling them to run a building-accreditation monopoly worth a million dollars a year.

Green was prosecuted twice for the same criminal charge, each time resulting in a hung jury.

Lennon's third deputy, Steve Kons, a former mayor of Burnie, is perhaps best known for drinking water contaminated with Atrazine after two people and their property in the north-east were accidentally sprayed with the herbicide by a Gunns Limited contractor in 2004. As water minister, Kons drank a glass of water from their poisoned bore to prove it was 'safe.' His fall from grace, however, was an incident now known as the 'Shreddergate Affair' in which shredded documents revealed he had lied to parliament. He resigned as deputy in 2008.

At the time, Lennon said Kons' decision to resign showed he was a man of strong character. There was 'always a way back,' he said, for people who admitted their mistakes. Never mind that Kons hadn't admitted anything until a former adviser fished the incriminating shredded documents out of the bin and passed them on to the Greens, who sticky-taped them back together and tabled them in parliament.

Even in retirement, Lennon has courted controversy. In 2009 he was called to give evidence against the island's top cop, who was accused of disclosing official secrets to Lennon and the police minister, Jim Cox. The leaked secrets were about none other than an investigation into Bryan Green and suspected impropriety in government appointments. When lawyers sought an affidavit from him, Lennon told them they couldn't meet at his house because the whole place was bugged. Instead, he suggested they meet at a shopping mall in Hobart's suburbs. They found him hiding behind a newspaper on a park bench and proceeded to Café 54, where they typed up Lennon's statement as he sipped a muggacino.

But in the words of Big Red himself, 'What's this got to do with forestry?'

Everything, I am told repeatedly by Tasmanian doctors, lawyers, activists, nurses, students, teachers, dole-bludgers, motel owners, carers, drunks, teenagers, writers, vets, waitresses, journalists, scientists, politicians, farmers, furniture designers, a man selling socks, foresters, gardeners, truckers, sawmillers – and the list goes on.

Forestry has to do with absolutely everything.

*

A local has given me a map of the island. Drawn in 1958 by a Cyril Lawrence, it is beautiful and childish. More fantasy than reality, it reminds me of a fun-park map; I half-expect sea monsters to be drawn in the surrounding waves. The map has divided Tasmania into five sections. In the north-west is a picture of a dairy cow, in the north-east a pick-axe and spade, in the west copper smelters, in the south-west apples, in the centre hydro-electricity, and in the south-east sheep. There is a sense of completion to this map, a satisfied loosening of the belt. As if to say, finally this place is organised.

I look at this map obsessively, but it takes a while to understand why I'm drawn to it. Then I realise what it is. The conflict over the forests may seem impossible to pin down but there is a certain stasis to the debate – and this map seems to symbolise it. For years the island's environmental movement has been locked in a much larger battle with the state and its various intimates – the Hydro-electricity Commission, the wool industry, British Tobacco, North Forests Products, the Forestry Commission, Federal Hotels, Forestry Tasmania and Gunns. Often these conflicts are framed as a choice between conservation and jobs, as though nature and economic progress are mortal enemies.

'I thought it was beautiful,' the then prime minister, John Howard, said in 2004. Just three days out from a federal election, he had spent the afternoon in the Styx Valley forest with Lennon. 'I said to a number of people,' he continued, 'I can understand why people who work in the forest industry get hooked on it.' When one reporter asked if he understood how people could get 'hooked' on protecting the same trees, Howard replied, 'I understand why people campaign for a whole lot of things, including job security.'

This familiar dichotomy makes resolution difficult, as struggling voters are told that the 'greenies' want to take away their livelihoods. Local politicians consistently spout working-class rhetoric, claiming to defend the 'common man' while cavorting with big companies and millionaires. A union job is increasingly used as a platform from which to launch a political career; one local joked to me that the CFMEU forestry spokesman, Scott McLean, who is oddly diplomatic when it comes to job losses at Gunns ('This is all the more reason why we need the pulp mill' is his regular response to job cuts) and is now looking to enter parliament, ought to use 'Vote for me. I know where you live' as his official slogan.

Lennon, who makes much of his working-class origins (particularly the fact that his parents once lived in the staff quarters of parliament house), is not averse to perks; he has been known to enjoy an upgrade to a $4000 room complete with its own butler at Melbourne's Crown Casino after brokering a multi-million-dollar internet gaming deal, and to commission expensive home renovations from one of Gunns' subsidiaries while simultaneously nursing the timber company's pulp mill through parliament. For a battler, he seems remarkably used to getting his own way.

The key players all seem more closely connected on the island; whereas on the mainland you might have to decipher seven or eight handshakes to get to the guts of a deal, here there may be only one or two. Far from making it easier to get to the bottom of things, this closeness makes things harder to untangle. Everyone wears and has worn so many hats, and two given people can be connected in a multitude of ways. The Greens senator Christine Milne has compared it to a game of musical chairs: 'They move around in Tasmania quite happily,' she said in 2005. 'The music stops and they leave the government and go on to the board of Gunns or they go out of the Forest Industries Association straight into the chair of the Secretary of the Tasmanian Department of Infrastructure, Energy and Resources – and they bring their secretary with them straight into the premier's office or the minister's office as an adviser and then move that person around. So you only have to pick up the phone once. There are no degrees of separation in Tasmania – pick up the phone once and you get them.'

Complicating things further, there is still a pattern in Tasmania of political offspring inheriting political seats. In the past two years, the late premier Jim Bacon's son, Scott, and his wife, Honey, have both run for election. The son of former Liberal premier Ray Groom was elected to parliament at the beginning of 2010 and until early this year, MP Michael Hodgman sat behind his son, Will, leader of the state Liberals; the Hodgman family has been in and out of parliament since the late 1800s. Paula Wriedt followed her father, Ken, into parliament, and Labor minister Alison Ritchie was pressured to resign last year after it was revealed she had employed her sisters, brother-in-law and even her mother as a 'speechwriter' on government money. 'For a fresh face with a name you can trust' was Carolynn Jamieson's motto when she ran for state parliament. She is the daughter of a former member of the island's upper house. Michael Field, former premier, is now the chair of Tasmania's Innovation Board, and is a good friend of the premier, David Bartlett. Field's brother Terry is one of Bartlett's advisors.

All this may well be a symptom of a small, insular island, and the inner circle of the environmental movement can appear equally tight-knit; Wilderness Society campaigners take jobs within the Greens Party and vice versa, and I am passed like a parcel between like-minded locals on both sides of the debate. Memories go back so far and alliances are so entrenched, the conflict can sometimes seem hopelessly static and repetitive.

There is, however, a point of instability: the Greens. Since their arrival on the political scene in the 1970s, their presence has upset the status quo both inside and outside parliament, and it is increasingly difficult for the major parties to dismiss them as a fad or a fringe movement. Their very existence poses a challenge to the cosy two-party system, and both Liberal and Labor seem to know it. Writing in the *Examiner* in the lead-up to the 2010 election, Robin Gray, a former Liberal premier and Gunns director known as 'the whispering bulldozer,' called on Labor and Liberal politicians to form a 'war cabinet' to prevent the Greens from holding the balance of power. Fighting words – so who is this enemy that has both sides so unsettled?

THE GREENS

B ob Brown peers at me, his blue eyes slightly misted as the steam from his tea curls upwards. He has a gentle assurance, perhaps a quality from his time as a doctor. On my journey I've met people who've described being 'Bob-struck,' while his political opponents have instantly soured at the sound of his name. There's no doubting there's something about Brown. Both a 'greenie' and gay, this quiet, studious man turned the island's status quo on its head when he entered parliament. People like him weren't even meant to exist here, let alone gain a massive following. And it was Brown's presence, followed by his fellow Greens politicians, that started to unpick what is still seen as a tight circle of private-school 'old boys,' their families, and union officials.

'When I got into parliament, I was so shy,' Brown tells me over a cup of tea and shortbread. We are sitting in his office, which seems to perch on the end of a pier in Salamanca. It was 1983 when Brown first took his seat. 'I thought it would be a good idea to go to the [parliamentary] Christmas party and as I was lining up for turkey, Michael Field [of the Labor Party] stood behind me and I said "Merry Christmas." His response was, "The only way for you is down."' Brown laughs at the memory, saying he didn't know what to do. Tongue-tied, he turned around, and the queue edged closer to the dead braised bird. 'Imagine his horror when he had to share the balance of power with us,' he muses.

For much of the island's democratic life, Labor has dominated state politics; both parties had settled into a comfortable routine of electoral spats and pragmatic coalitions. When the Greens arrived on the scene, it was as if all those pacts and favours of the past, the cutting of red tape for 'mates,' were under threat. The hostility towards the newcomers was palpable. When Brown took to walking the corridors of Parliament House, turning off lights in empty offices, ministers responded by sticky-taping them permanently on. When Christine Milne and Di Hollister were elected as the first female Greens MPs, a Liberal frontbencher called them 'political sluts' after their maiden speeches. But although the Greens were ridiculed, ignored and undermined, the votes kept coming.

Fury at the Greens' existence has never been limited to the state parliament: the vitriol has seeped into the wider public domain, fed by outlandish political statements. In the late '70s and early '80s, the Greens made the damming of the Franklin River a key issue in the federal election. The state Liberal leader, Robin Gray, donned boxing gloves and shadowboxed down the main street of Queenstown, declaring he would fight any opposition to the dam. A week later, perhaps taking Gray's call to arms too literally, four Queenstown men bashed Bob Brown over the head with a wheel brace.

On another occasion, when Brown tried to approach a gathering of Hydro workers, Labor ministers and the HEC chief out in the field, workers drew a circle around him in the dirt. If he stepped outside the circle, they said, they'd beat his brains out. The Labor MPs and HEC chief went on to share scones, jam and tea with the workers, effectively sanctioning their behaviour. Years later, when a man shot at Brown's head, police reluctantly charged the shooter with 'discharging a gun on a Sunday.' No sense of right or wrong seemed to prevail; no Labor or Liberal politician came forward to temper this bizarre open season on environmentalists.

*

According to former Labor premier Michael Field, the Greens did more than challenge the cronyism of the island's parliament; they also gutted the heart and soul of the Labor Party. 'The electoral imperative of the Greens is to diminish us,' he says to me in a Hobart pub. 'We're battling over the same turf.'

But environmentalists only formed a party to fight the flooding of Lake Pedder in 1972 because both existing political parties had declared Pedder a 'non-issue.' When I point this out to Field, he concedes and blames Labor's old guard for wasting the opportunity to embrace these new voters. 'They had built themselves on a "jobs for the Aussie battler" agenda,' he says. 'This new "post-materialist" voter was alien to them.'

When, almost a decade after the flooding of Lake Pedder, the issue of protecting the Franklin arose, this 'old guard' dug their heels in even further, causing its more progressive ministers to reconsider their loyalties. Sensing an internal rift, Liberal leader Robin Gray played the first of a series of strategic hands. Originally from Victoria, Gray managed to convince his own team to stand united for the damming of the Franklin in order to gain power in parliament in 1982. He helped form the pro-dam group Organisation for Tasmanian Development, which was responsible for bumper stickers such as 'Doze in a Greenie: Help Fertilise the South-West,' 'If It's Brown, Flush It Down' and 'Keep Warm This Winter:

Burn a Greenie.' The strategy worked. While the wild river was saved by a High Court decision, Gray's shrewd pro-development stance saw him hold onto the premiership for seven years. In 1985, the Gray government was part of a national lobbying campaign for the federal government to renew woodchip export licences for the next fifteen years. The woodchippers got the all-clear and the state Greens prepared themselves for the onslaught against Tasmania's forests.

It was a pulp mill that brought about Gray's end as premier. A local timber group, North Broken Hill, in partnership with Noranda, a Canadian company, proposed the Wesley Vale mill. Gray was pushing hard for it, despite growing opposition not only from the Greens, but from farmers, graziers and fishermen. Then, on North Broken Hill letterhead, the premier recalled parliament to pass an act to enable the Wesley Vale mill to be built. He was blatantly allying the government with the corporation in the face of the health and business concerns of the electorate. He lost the next election.

When the Field Labor government won power in 1989, it was bittersweet. There was the unforeseen matter of a $4 billion state bankruptcy and he had to share the premiership ... with the Greens. Robin Gray had lost the election by one seat, and the Greens had recorded their strongest support ever, winning five seats and the balance of power. Tasmania became the second place in the world after Germany to be governed by a red–green alliance.

In response, Gray insisted on a fresh election. So too did a group calling themselves the 'Concerned Citizens for Tasmania.' They placed full-page advertisements in all the state newspapers and started gathering a petition. Around this time, one of Field's ministers, Jim Cox, received a phone call from a public telephone booth in Melbourne. The minister was offered $110,000 to cross the floor and collapse the accord. Cox alerted Field and the police. A sting operation was set up and the caller was arrested. He was a radio-station salesman who worked for Edmund Rouse, Tasmania's media magnate, then chairman of Examiner-Northern Television which had a substantial holding in Gunns, then known as Gunns Kilndried Timber. Rouse was also a chairman of the timber company and he had hired the radio-station salesman to coax the minister across the floor.

Rouse was to spend eighteen months in prison. In his defence, his lawyer said he had believed he was doing Tasmania a great favour, that the bribe was a gift to the island's future. In other words, the Greens would ruin Tasmania. But Rouse wasn't known for his generosity: former employees bitterly recall him visiting the newspaper before Christmas and handing out chupa-chups in place of a bonus or staff gift. A Royal Commission into the affair revealed that Rouse feared he stood to lose between $10 million and $15 million, in part through a potential cap on woodchip exports and the protection of swathes of forest from logging.

The commission also revealed Robin Gray had $10,000 stashed in freezer bags in his home, a pre-election donation from Rouse. The former premier admitted Rouse had told him about a potential Labor defector but denied knowing of any attempt to bribe the minister. It also appeared that Paul Lennon, then secretary of the Tasmanian Trades and Labour Council, was in Gray's office when names of potential defectors were tossed around. In an initial statement Lennon said as much, but then denied it, saying the lawyers taking his statement must have 'misheard' him. Also under scrutiny were two Gunns directors, David McQuestin (also managing director of Rouse's company, Examiner-Northern Television) and John Gay. The latter was cleared of any wrongdoing, but David McQuestin's behaviour was found improper and in breach of commercial morality. Although Gray's behaviour was found by the commission to be evasive and dishonest, he stayed in parliament until he was eligible for his full superannuation payout. When he finally resigned, there was a seat waiting for him at the table of Gunns' directors.

And the 'Concerned Citizens for Tasmania' campaign? Turns out that was the fictional brainchild of yours truly, Robin Gray.

*

A former schoolteacher from Railton, Michael Field is a typical Labor politician. A stocky man with peppery white hair, he tells a few stories with some gregarious thigh-slapping, throws a beer back and drops in and out of Australian political gobbledy-speak. He classifies people as 'quality of lifers,' 'traditionalists,' 'post-materialists,' 'latte-sippers' and the working class. He went into union politics in 1976 after a run-in with the police soured his teaching rounds. 'They caught me with a couple of marijuana seeds growing in half a beer can,' he says laughing. I meet with him a few times, twice in a Hobart pub and again at a café during a busy Hobart lunch hour. At one point during our first meeting, Bob Brown and his colleagues walk past our table. There is an awkward rush of greetings, steely handshakes and quick getaways. I assume this is just how things are in small cities, but when I meet Brown and mention the run-in, his eyes widen. 'I was gobsmacked. I haven't seen Fieldy in years.'

'It was a forced marriage,' is how Field remembers the first Labor–Green accord. 'They are difficult people. If one of them tries to do something, the others will call them a sell-out. It was impossible to get things done.' Others remember the time differently, claiming that Labor signed the accord, gave the Greens a couple of bones and then governed alone, ignoring them if they piped up in pre-cabinet meetings. But if the accord was a forced marriage, neither partner was aware of the paltry dowry they were about to inherit.

'When I opened the books, I was furious,' Field says to me. 'I sent Rouse the budget papers while he was in jail, so he could look over the numbers and see the train wreck that was the government under Gray. I was so offended that he wanted to bribe someone to stop me from being premier when Gray had run the state into the ground.' He pauses, takes a sip of beer, and continues. 'Rouse wrote back and you know what he said? "Well, rest assured, there's one person worse off than you and me, and that's Christine Milne's husband."' Slapping his palm on the table, Field laughs loudly. 'Christine's husband! Ha!'

But with or without Rouse, the island's parliamentary process was, and many say still is, heavily dependent on personal relationships, and the Greens, well, they don't have mates. Instead, according to their opponents, they have ideological positions, from which they will not budge. 'If you want to express beliefs, then go to a church, not government,' says Field, a little petulantly. Under the Labor–Green accord, the size of the World Heritage reserve doubled. But that wasn't enough, he says. 'It's never enough for the Greens.' The divorce was bitter – 'acrimonious.' Field introduced legislation that would allow an increase in woodchipping. It was a blatant breach of the accord and the one thing he knew the Greens couldn't agree to: their whole platform was built on the protection of old-growth forests. If they agreed to the legislation, they'd no longer be the Greens.

However, when Field pushed for the bill and the Greens moved a successful no-confidence motion, he wavered, not wanting to lose his premiership, and withdrew the legislation. The bill was dead in the water. But one Labor minister – David Llewellyn – wouldn't let it go. Backed by Paul Lennon (now on the Labor backbench) and the timber industry, Llewellyn lobbied for the bill's reintroduction, knowing it would bring down the accord, Field as Labor leader and the cabinet with it. Concurrently, Jim Bacon, then union secretary having replaced Lennon at the head of the Tasmanian Trades and Labour Council, cut a deal with who else but Robin Gray, who guaranteed safe passage of the bill from the Liberal opposition before letting the house of cards collapse. When the bill's passage was secured, the Greens refused to continue their support of Labor and Field was forced to call an election. The Liberals appointed Ray Groom – an AFL footballer – as their leader, fearing the stigma around Gray, and won power. One of the first things they did was introduce anti-protest laws: forest activists could now be fined up to $20,000 and sentenced to jail for trespass.

THE BACON YEARS

The Liberals might have ousted Labor, but after four years of leadership they too became dependent on the support of the Greens, who in 1996 won back the balance of power following a public backlash against the Liberals' decision to raise politicians' wages by 40 per cent. Two years later (and two years early) another state election was called and Labor, now led by Jim Bacon, returned to government.

A moustachioed charmer, Bacon was a master of reinvention. The former Melbourne private-school boy was a devout Maoist at university, then an official with Victoria's notorious Builders Labourers Federation, working closely with its legendary boss Norm Gallagher. In 1986, Gallagher was jailed and the BLF was deregistered after a commission found he received more than $150,000 in bribes in labour and materials to build a beach house. Bacon moved on, unscathed. In Tasmania he met Honey, a croupier and the public face of Australia's first legal casino. They married and in 1989 Bacon became secretary of the Tasmanian Trades and Labour Council, taking over from Paul Lennon, who had taken a seat in parliament. The new union secretary had no historical affinity with forestry, but did have a natural affinity with power. Having married the public face of what is still thought to be one of the most powerful corporate interests in Tasmania, Federal Hotels, it was only fitting that he should now set his sights on the timber industry, its most powerful. After helping to engineer the collapse of the first Labor–Green accord in 1992, he followed Lennon into parliament in 1996, and became leader of the opposition the following year.

Tasmania operates under a unique electoral system known as Hare-Clark, which by the late 1990s had enabled the Greens to become a powerful political minority. In the mainland states, each electorate is represented by a single member in the lower house; there can be only one winner per electorate, regardless how many votes were directed at other candidates. In Tasmania, however, each electorate is represented by five members in the lower house, who are elected through a system of preferential voting. This system of proportional representation aims to more accurately reflect the votes cast in each electorate. In 1998, while still in opposition, Bacon's Labor put forward a plan to reduce the number of seats in both houses, thereby restoring the cosy two-party system the old guard had become so accustomed to. The Liberal government had recently introduced legislation to raise politicians' wages by 40 per cent (a pay rise Labor happily accepted). If the number of seats in parliament was reduced – so went Bacon's sales pitch – the wage hike would not impose an extra cost on taxpayers. Getting rid of the Greens was just a fringe benefit.

The Liberal government, reliant as it was on a fragile agreement with the Greens, did not initially support the change, although one Liberal backbencher, Bob Cheek, crossed the floor to vote for it. In his post-politics memoir, Cheek recalls that on the evening before the motion, Robin Gray, by then a director at Gunns, left a message on his answering machine. In his distinctive gravelly voice, repeating himself until the tape ran out, Gray said, 'Get rid of the Greens,

Bob. Just get rid of the Greens and do Tassie a favour. Don't back down, we need you ...'

The legislation didn't make it through the first time, but it did the next. In the process, it collapsed the Liberal–Green accord and marginalised the Greens yet again. Under the new system, none of the seats held by the Greens existed. Christine Milne said she wept the night the bill was passed. When an election was called a few months later, only one Greens member, Peg Putt, secured a seat. On her first day back in parliament she dragged a deckchair into the aisle between the two main tiers of seats, announcing it was the only part of the house that wasn't hostile.

*

As premier, Bacon was populist and pro-business. With Paul Lennon and David Llewellyn at his side, he ran the most pro-wood-chipping cabinet the island had ever seen. The year before his premiership, the state Liberals had signed off on the 1997 Regional Forest Agreement, a twenty-year plan for the management of the state's forests. The Greens, led by Christine Milne, had serious concerns about the RFA, but were keenly aware that if it didn't get through, a Labor version would be much worse. In opposition, Bacon had regularly goaded the minority Liberal government for putting too much land into reserves and being under the Greens' thumb.

It was often observed that Bacon and Lennon were a particularly clever combination; Bacon played the role of salesman and Lennon of the brutish and uncompromising enforcer. It was a good time to be premier. Tasmania was finally lifting itself out of the Gray government's debt, and Bacon became the king of feel-good politics. Hugely popular, he restored confidence to a Tasmania shaken by a string of hung parliaments and bad press, which to Bacon's mind was mostly the fault of the Greens. But his successful campaign to shrink the parliament had dealt with that.

In the seven years of Bacon's premiership, the island dominated not only Australia's woodchip production but also the world's hardwood-chip export market. The RFA removed limits on native woodchipping, essentially giving the state government the right to decide the fate of the native forests for the next twenty years. Gunns took full advantage, producing record profits.

On the surface, however, Bacon portrayed himself as an 'arts and culture' kind of man, setting up Ten Days on the Island, an arts and literary festival, as well as Tasmania Together, a 'listening committee' set up to draft a blueprint for Tasmania's future through community consultation. He was 'pro-jobs,' which was a clever way of being 'pro-business,' but not necessarily 'pro-government business' – he oversaw the giving away, for free, of a fifteen-year gaming licence worth over $120 million to Federal Hotels.

Bacon had a peculiar approach to his 'feel-good' cultural investments, as if they were inducements to the public to behave rather than normal things a government might do for its citizens. When a group of artists presented him with an open letter criticising the logging of old-growth forests, he is said to have slammed the petition on his desk, saying, 'What do you people want? I gave you a festival.' Later, when it was announced that Forestry Tasmania would be a major sponsor of the festival, local writers and artists boycotted the event and raised $75,000, not only matching but exceeding the controversial sponsorship by $25,000. When the festival board rejected their money, leading writers such as Richard Flanagan and Tim Winton withdrew their work from the festival's literary prize. Bacon dismissed them as 'cultural fascists.'

The Tasmania Together forum also didn't turn out so well. Kicked off in 2000, the community leaders found an overwhelming number of locals wanted old-growth logging to end. A government-sponsored survey found that 72 per cent of citizens were in favour of protecting all old-growth forest. They concluded that the island needed a plan to stop logging high-conservation-value forests by 2003. But their report was stonewalled and two of the community leaders, Anna Pafitis and Gerard Castles, were pushed out of the committee after objecting to the government's interference. In his 2007 essay for the *Monthly*, 'Gunns: Out of Control,' Flanagan recalls meeting a state politician shortly after Castles had published an opinion piece describing the undermining of the consultative process. At the mere mention of Castles' name, Flanagan writes, the politician flew into a rage. '"The fucking little cunt is finished," he said in front of me and my twelve-year-old daughter. "He will never work here again."'

Then in 2004, after a formidable premiership, Jim Bacon announced he had inoperable lung cancer and let go of the reins. His loyal treasurer, David Crean, ill with kidney disease, also left politics. Lennon was alone at the helm. With a drastically reduced parliament of twenty-five members, he had fewer ministers to rely on, and his opponents smelt blood. Big Red was wounded. He appointed David Llewellyn as his deputy but the dynamic had permanently shifted. 'I never wanted to be premier,' Lennon says when I ask if he'd ever considered the job before Bacon's illness. 'It was a hard time. Don't think I've fully recovered from it. There was three of us, and then one morning, there was just me.'

Within four months, Bacon was dead. The eulogies ran thick and fast. Paul Lennon gave up the smokes and Robin Gray applauded Bacon for 'turning Tasmania around,' never mentioning that it was he who had sent the island's finances backwards, bequeathing an enormous debt that had paralysed the state for almost a decade.

BANANA REPUBLIC

T he art of politics, say many, is the art of compromise, also known as cutting deals. Australia's former minority party, the Democrats, were impaled on the sharp end of a compromise in 1999. At a crossroads, the Democrats had the option to partner with the up-and-coming Greens or to broaden their membership by embracing a more flexible ideological platform; in other words, negotiate and water down their ideals. When the Democrats chose the latter and supported the Liberal government's GST tax package, two of its senators crossed the floor (reflecting a bitter internal feud) and the plug was pulled on their support base. That is the danger of trying to play politics while remaining on the moral high ground.

How well have the Greens negotiated this path? They have built their popularity on claiming the moral high ground, but they now face a classic political dilemma. Stay true to one's principles and hold out for reform, even if it means staying perpetually on the fringes? Or shake hands and make incremental changes, knowing every deal done has the potential to lead you astray?

'They wouldn't budge,' Michael Field says of his Labor government's partnership with the Greens. 'Bob Brown doesn't trust people. My pop psychology says that would come from being gay in a family of cops and brought up in a small town.' But Field is being sneaky when he says the Greens were, and are, impossible to work with. Not counting the current parliament, they have held the balance of power twice, once with Labor and once with the Liberals, and both times the Greens brought down these minority governments by voting against their alliance partners. But what Field fails to mention is that both cases involved bills designed deliberately to break the alliance. In the case of the first Labor–Green accord, it was the expansion of wood-chipping; in the Liberal–Green agreement, it was the shrinking of parliament, with the clear intention of disempowering the Greens. These were never potential compromises – they were always going to end the accords. And between these two fallings out, the Greens, under Christine Milne, partnered with Liberal premier Tony Rundle to achieve gun law reform, an apology to the island's stolen Indigenous children and the legalisation of homosexuality. They even compromised on the forests, supporting the 1997 Regional Forest Agreement, despite its environmental flaws and heavy allowances made to industry, simply because the alternative was much worse.

Environmentalists have been forced to play this kind of game for decades, compromising to achieve a 'less bad,' if not a terribly good, outcome.

'We're sick of forests being used as bargaining chips,' a local tour operator said of recent state plans to push a new road through the Tarkine rainforest in exchange for the reservation of 653 hectares of high-value forest scheduled for logging. 'If the forest is of equal value to the virgin rainforest they plan on clearing, then it ought to be protected regardless.'

Ecologists and biologists all over the world are realising that this sort of exchange – the 'you can have this while we demolish that' approach to preservation – is not working. Such swaps are often driven by short-term concerns, thrown together to appease voters before an election, and can end up causing unforeseen damage: cornering species in a shrunken habitat, dramatically reducing food sources and ignoring patterns of migration. Wild places link together, and these corridors need to be restored.

'Ultimately, I don't want to see reserves at all,' says Pete Hay, the author of *Main Currents in Western Environmental Thought* and a former Labor adviser. 'I want us to become a mature enough species that we will not pose a threat to wild processes. At the moment we lock wilderness areas up for their own protection, as if they are some sort of idiot child that can't be let loose on the streets because the nasty kids are going to bully them.'

For the moment, however, this is the system in which conservationists must work; deals must be done and compromises made. But can genuine negotiations take place in a political culture as polarised and matey as Tasmania's? Although it has long been touted as a virtue, I'm beginning to wonder if solidarity isn't one of the most destructive social forces on the island. Does unity inevitably lead to dogma, and to hostility to anyone from the outside?

An inflexible code of unity is prevalent on both sides of the forest conflict. When a group of local timber workers spoke out against woodchipping, the chairman of the Forestry Industries Association of Tasmania accused them of 'pissing on the tent from the inside.' Jimmy Barnes, whose 'Working Class Man' is a union anthem, was declared a 'working-class traitor' when he came out on the side of the old-growth forests.

And of course, there's Peter Garrett, pilloried as a 'sell out' by the activists and served his balls on a plate by Ula. On the car wrecks at their blockades, they've scrawled altered lyrics from his songs;

I spy 'How can you sleep when our old growth is burning?' and on another, simply, 'Will the real Peter Garrett stand up?'

The thing about solidarity, of course, is that people expect it to go both ways – having stuck loyally with Gunns, the island's timber workers expect some support in return. In 2005 a former log-truck driver from the south of the island, Gary Coad, told a local news crew he and his colleagues felt entitled to industry support. 'We came up and fought for John Gay's livelihood. Well, now it's time for him to turn around and do the same for us.' Coad meant it literally; the previous year he had swung his ute into the emergency lane of a main road and jumped out to beat up a local resident who was standing there with a video-camera waiting to film the log trucks coming out of the valley. The last shot recorded a glimpse of the attacker's furious face, his fists balled up.

Money and power often seem to inspire the strongest sense of solidarity, and these ties can weaken other bonds, including Labor's professed commitment to the working class. Graeme Sturges, at the time a Labor minister, was once stopped by a security guard at the door of a function. 'Don't you know who I am?' he yelled. 'I'll have your fucking job!' Sometimes these economic alliances are predictable, as when Gunns boss John Gay and prime minister John Howard – leader of a vocally pro-business party – presented a united front in Launceston during the 2004 federal election. Others are more surprising – consider that a sea of union-clad timber workers, a quintessential Labor crowd, cheered Howard and Gay on.

Is it really as simple as money talking – a case of one powerful company running the state? Activists keep telling me that Tasmania is a banana republic, completely beholden to a single, unsustainable industry. I get tired of hearing this line; it seems so simplistic, and so resigned.

'What about a revolution, then?' I ask one local who repeats the banana republic claim. 'Don't you need one of those?'

He laughs before looking at me sharply. 'What makes you think there isn't one happening here?'

I look around at the sleepy town with one main street, lush green paddocks, the odd person walking their dog, mutt and man both breathing puffs of cold air. Nope, I think. No revolution here.

GROUNDSWELL

THE FORESTER

P aul Lennon had laughed at me when I wondered out loud why there seems to be so many Tasmanians doing so much activism for no money and little, if any, recognition.

'You don't really believe that do you?' he said. 'They're being paid. They're professional protesters.' Mistakenly I think he is referring to the Greens politicians, the four, sometimes five, full-time employees of the Tasmanian Wilderness Society or the three employees at Environment Tasmania. But then he added, 'You don't think people who abseil down buildings aren't trained? They're paid.'

Then it was my turn to laugh, thinking about the activists at Camp Florentine, scrounging around at the tip shop for spades or an axe-handle, their hands wriggling down the back of car seats for coins to pay for petrol, dumpster-diving for food. I think especially about their recent 'skill-share' day where they gathered under a tarp, sitting on damp cushions and cold rocks for workshops run by law students, while others taught new arrivals how to climb, connect cables and spot for one another. A sharp look from Lennon shut me up.

'Well, they've got you fooled if you believe they're doing this for free.'

But it isn't just the activists at Camp Florentine who have dedicated their time to fighting the forestry industry. There might not be a revolution brewing, but the more time I spend on the island, the more people I meet who have refused to follow the conflict's rules. These people aren't ratbags or ferals, or even passionately 'green,' but they've spotted something odd and said something. In a sense the island reminds me of a little America, deeply conservative, but with the potential for radical spurts of steam to erupt should anyone try to rule with too firm a hand.

*

I ask Bill Manning why he didn't just quietly accept how things were and go home at 5 p.m. like everyone else. He shrugs. 'You think you've this tide of support behind you, everyone's nodding and agreeing with you, but then as soon as you step out – nothing. You're on your own.' He laughs. 'I should have known from an incident when I was eighteen. I was at an eighteenth birthday party and there was this banging on the door. The parents answered it and it was a bunch of bikies from across the road adamant that one of us had bashed one of their own. The parents didn't know what to do, so they just closed the door. Next thing, their front lawn was covered with bikies and me being me, I yelled out to the others "Let's get 'em, boys!" and charged outside, only for the front door to close behind me. Last thing I remember was one fella swinging a fence paling at my head.' I start laughing as Manning shakes his head. 'I should have learnt my lesson then.'

Manning lives with his wife in what was once their weekend beach shack. Gruff and wary, the former state forester is flanked by his black and white border collies when he greets me at the front door. One dog is stiff and shuffling, while the other is determined to monopolise my ball-throwing skills. Standing between them, Manning himself seems to have a bit of each quality. With rough white sandy hair, a surfer's face licked by the salt and sun and thick calves, he is young at heart but somewhat battered.

In 2003 Manning became Tasmania's first forestry insider to allege broad-scale illegal destruction of the island's forests. By then he had been employed as a forester for thirty-two years, stubbornly holding on to his idea of what a forester was meant to be as the landscape around him – both physical and bureaucratic – changed its colours. His last ten years in the job shattered him and convinced him to blow the whistle on what he says is a ruthless industry rife with 'bullying, cronyism, secrecy and lies.' When he was subpoenaed to testify at a federal Senate enquiry into the plantation industry, he slammed the forest industry's ability to regulate itself.

> The forest industry has become so woefully negligent in its practices that it has been forced to be exempted from all other state environmental, planning and land management legislation for the simple reason that were it to be judged by the legislation that other Tasmanians have to abide by, it would be found to be comprehensively in breach of Tasmanian law … science has largely been ignored due to the influence and dominance of the woodchip industry foresters on the Forest Practices Board and the Forest Practices Advisory Council. The erosion of best practice has been compounded by the self-regulation of the industry, which has been so ineffectual as to render it virtually non-existent. This has meant that standards of forest practice have actually dropped markedly and the industry is in virtual regulatory free fall.

At the time, Manning predicted that once he had given his evidence in Canberra, his career and perhaps even life as he knew it would be, quite simply, finished. 'And then I am done,' he said. 'Like a dinner.'

Paul Lennon, then deputy premier and forests minister, had ordered that no representative of the Forest Practices Board was to appear before the Senate committee. Manning's evidence and allegations were left in limbo, never to be adequately addressed. Before the Senate hearing, when Manning attempted to convey his concerns about the Forest Practices Board to the attorney-general, his evidence was ignored for six months. When it was finally considered, the ombudsman handed it to the secretary of the Department of Infrastructure, Energy and Resources – who was previously the head of the Forest Industries Association of Tasmania and now reporting to Lennon.

Instead of investigating his allegations, much was made of Bill Manning's clinical depression. 'They all said I had a mental disorder when I blew the whistle, that I was mentally unstable,' Manning recalls. 'Wilkinson [his boss], Lennon, they all came out and said I was ill.' The forester had never denied his illness and had made it very clear in his evidence. But it was a question of which came first – the mental illness and then the bullying, or the bullying and then the mental illness?

Manning has seen a lot of logging in his time. In the early '70s he did his technical forester's apprenticeship, something like a cadetship: the young foresters were rotated around the state, doing month-long stints in various areas of forest management. 'You get involved in strange things being a forester. You fight fires with farmers, take turns to keep watch for embers, help deliver calves in the middle of the night. We worked hard in those early days, drank hard, fished and chopped wood for our fires.'

In Maydena, he worked with the senior ranger in National Park and was appointed to oversee one of the many logging crews that came through the area in the mad rush to salvage timber from around Lake Gordon before it was flooded for the hydro-electrical project. It was 1973 and Manning, like many urban 22-year-olds of the time, had long hair. A man by the name of Jim Hall was head of the crew he was to work with. 'He was a giant of a man. Ex-cop and ran his own logging and sawmilling company.' There wasn't enough time to cart all the sawlogs out of the valley, so the logging crews started to stack them on islands above flood level. It was Manning's job to mark out roads to these islands for Hall and his men to bulldoze. 'He'd told me not to bother with lunch, that he'd bring enough for everyone.' Lunch was a bottle of scotch passed around. 'So I was leaning against this stump, pretty exhausted, and Jim came up behind me, grabbed my hair and cut it off with his axe. He didn't scalp me, but it was pretty bloody close.'

A few weeks later, when Manning's time to move on was imminent, Hall asked him to stay on and work for him. By surviving the macho rites of the backwoods, Manning had earned respect. 'No way did I take the job,' he laughs. 'Not something you'd expect the managing director of a logging company would do to a state servant, eh?'

By the time Manning completed his apprenticeship, the clear-fellers were becoming voracious. He had seen what they did to Lake Gordon down in the south-west, and this was tenfold. 'They were picking the eyes out of the place.' His training seemed to be at complete odds with the timber industry's changing priorities: 'My values were that of the old foresters, which was jealously guard your patch from companies and contractors. Look after the forest and it looks after you.' But as traditional logging crews were being amalgamated with the woodchip crews and pulp quotas taking precedence over sawlogs, Manning's role was becoming increasingly woodchip-focused.

In 1975, APPM, a paper and pulp company based in Burnie, hired a professional forester to oversee its logging practices. 'It was a first for Associated Pulp and Paper Mills, and for any of the companies actually. We got on really well.' The two foresters could see the dire consequences of chipping the industry's finite resource and started to formulate a landscape-management plan to retain potential sawlogs for the future. 'But the manager of the Tamar woodchip mill got wind of our plan and wrote to us to say that under the Wesley Vale Act anything that was not a sawlog then and there was to be pulp for the chippers. I was gutted, because he was right. The Act actually did outline this.' Unable to pursue their plan, Manning decided to mention the predicament to the sawmillers he worked with, knowing they'd be outraged. 'They kicked up a fuss and took it to parliament. It became an issue and a group of ministers organised a meeting with us at the Cluen Tier.'

For Manning, this is when things began to sour. Before the ministers' arrival, he was pulled aside by his boss in the Forestry Commission (now Forestry Tasmania). 'I was told that if I didn't want to spend the rest of my life posted in Queenstown, I was to shut my mouth.' Taken aback, Manning did as he was told. 'The ministers came, asked us questions and I gave them nothing. They knew something was up, but no one said anything. That's when I started to realise my job wasn't about forest management.' Eventually, parliament lost interest in the debate, and APPM sacked their forester.

In the early '80s, Manning went on to become the deputy district forester for the Launceston region. At coupes he could access, he sorted the felled trees into sawlogs, woodchips and specialty timbers.

Realising that much of the wood outside his supervision, regardless of its merits, was being siphoned in a steady stream to the chipper, he started positioning his car at the gates of woodchip mills and inspecting the trucks driving in. 'I stopped trucks with sawlogs on their tray and turned them around back to the bush to unload and start again.' When workers argued with him, he painted the outline of where the wood could be sawn on the trunk with a spray can. Angry that their shortcuts were being foiled and under pressure from their bosses, the truckies began to warn one another on their two-way radios when they saw his car. For the embattled forester, it was exhausting work.

Disillusioned, Manning took leave from the Forestry Commission with the hope he wouldn't have to return. Joining a diving crew, he collected abalone on the west coast. 'We'd go out in this little boat, wait for the weather to go off and the water to flatten out, so we could dive.' It was the late '80s and he was earning good money. 'There were lots of parties and diving and camping out on the beach.' Just when he was almost certain he wouldn't have to return to the Commission, Manning got the bends. 'We were diving in a rock pool and I didn't think to add weight to my diving belt.'

The process of bobbing up and down doubled the nitrogen in his body and he ended up in a decompression chamber, the doctors lowering his body down to a pressure depth of 30 metres and bringing him slowly back up to the surface pressure. With a texta they marked out on his skin where he felt pinpricks and loss of feeling. 'My diving days were over and I was slightly brain-damaged,' says Manning. 'I had to retrain my brain.' When he was ready, he returned to the Commission, where he was put on roads and surveying work. 'It was slow and boring work. But eventually my brain came good.' When the position of forest practices officer came up at the newly formed Forest Practices Board, an independent governing body the timber industry had agreed to accept, Manning jumped at it. 'I was the only applicant for the job. I was stupid. I believed it when they promised me six staff. I got three for a little while and then they disappeared and it was just me for twelve years.'

All forestry operations in Tasmania – including the harvesting, seeding and burning of forest, establishing plantations, the construction of roads and quarries – were now to be regulated by the Forest Practices Board (since renamed the Forest Practices Authority). Manning's role was to be an on-the-ground regulator of the new system. Each coupe was to be assigned a forest practices officer, who was to inspect the coupe prior to clearing, draw up a sustainable logging plan, have it signed off by another officer and then supervise the logging crew. Manning soon realised he was one of just three forest practices officers employed independently by the board and the only officer to go out into the field. The rest of the officers, about 120 in total, were (and still are) employed by the logging companies themselves. A system of 'co-regulation,' Lennon called it.

Manning was assigned to supervise logging operations on private land. He discovered private landowners were being left with a slurry of useless land after loggers had been through and many were devastated – emotionally and financially. At the handful of the private coupes he could supervise, Manning took his supervision seriously. To begin with, his strict adherence to the forestry code enraged many of the contractors. Loggers rattled boxes of matches at him, saying they knew where he and his family lived. Trees were pushed across logging roads on his way out, so he would have to leave his car and walk out to the road. It was hard, lonely work but Manning eventually started to win over the workers by setting up incentives. 'I let it be known that the most trustworthy workers got the best coupes with the most volumes. It gave them an incentive to do a good job and it became a kind of competition among the men. They used to ask me about one another – what did he get on his coupe, that sort of thing.'

At one point Manning and a group of timber workers were sent to Victoria to observe a new system that had been implemented there. 'Timber workers were given a number of points, like demerit points on a driver's licence, and if they lost all their points, they were out of the game, so over time they would get rid of all the idiots and cowboys in the industry. Our group liked it unanimously. But when we tried to take the system back to Tasmania, there was an uproar.' The timber industry and the forestry minister 'didn't want police in their forests.' 'We were directed to review our decision over and over until we agreed with them. I pushed and I pushed and all it did was bury me,' he says. In 1998 he had a meltdown.

After taking sick leave, he was diagnosed with depression. In hindsight the diagnosis makes sense to Manning. 'I know now that I was exhausted from constantly doing my job and getting told not to do it. I was overworked, bullied, threatened and never had a single win. But then, I had no idea what was wrong with me.'

When Manning returned to work, he was transferred from his position overseeing private timber operations and into state forest jurisdiction. He was sure his new role assessing state coupes after they had been logged would be easier. Naively, he tells me, he believed coupes run by Forestry Tasmania would be managed responsibly. When he saw how they were dealt with, he recalls, 'I was stunned.'

'The whole time I was trying to supervise private timber operations,' he says, 'I'd taken solace that at least things were managed better in state forests.' When he began his new desk job, however, he found a different story. Of all the forest plans submitted, he was required to audit a random 15 per cent. He discovered that the Forest Practices Authority received only the cover sheets of the twenty-page plans, with the full text held by the supervising officer, who was almost always employed by the logging company involved (usually Forestry Tasmania). These officers would draw up a plan for each coupe, certify it and give their word that the logging complied with the rules. Manning had to ask the officer for the full plan, often four or five times. The documents could easily have been altered retrospectively, but still he found numerous breaches of the code. 'There was a huge negligence,' says Manning. When he filed his report, the chief forest practices officer, Graham Wilkinson, asked him to redo the audits. 'He then got a second person to do them to show me how the results could be re-interpreted, but I still wouldn't change my findings.' Manning's audit had found nearly 100 breaches of the act, but Forestry Tasmania was prosecuted for none of them.

To Manning it was glaringly obvious that the logging plans had been drawn up and adjusted to maximise the area for woodchipping. He laughs cynically. 'I knew how it was going to pan out by then, so I was prepared. For every coupe they challenged, I'd made sure I had on hand a specialist's report backing my findings – so it all had to go into the annual report, even if Forestry Tasmania wasn't prosecuted. So then I got put onto the fauna audits, then regeneration surveys. I began to joke that I'll be cleaning toilets soon.' He was aware they were trying to move him on. 'They began to override my breaches without even visiting the coupes.'

Manning and Graham Wilkinson no longer spoke and communicated by email only. 'His secretary would call me when I was out in the field and say Graham's sent you an email can you respond to it in the next hour – and I'll be like, "I'm in Smithton," and she'd say, "Are you refusing to do this?" Basically they were trying to get enough crosses against me to fire me.'

Then, in 2002, Manning issued a breach after discovering debris had been bulldozed directly into a stream. When he spoke to the district supervisor and said he was going to issue the crew with a fine and a notice to remove the debris, the officer's reply surprised him.

'Please do,' he said. 'It wasn't under our instruction. This was Forestry Tasmania's policy.'

It was to be the last notice Manning would issue. In response to a directive that he withdraw the penalty, he said that if the Forest Practices Authority didn't act quickly on breaches such as this, they would oversee a complete breakdown of self-regulation. Within eight days Manning's charge books were taken from him and he was shifted to a different department. When Paul Lennon was queried about Manning's evidence in parliament, he said the forester wanted a rigid and punitive approach whereas the Forest Practices Authority 'takes a different attitude, preferring a carrot-and-stick approach, working with the industry to effect improvements to the system.'

*

From Manning's deck, overlooking his red '75 Holden ute and vegetable patch, we can see the ocean, close enough to hear the waves slapping the shore. Inside there are binoculars at the ready to catch a glimpse of migrating whales and dolphin pods. Manning sits with his collies and tries to summon the desire to surf. He says he hasn't caught a wave in years. He's had depression for twelve years now. 'I'd eat horseshit if I knew it would cure me.' He put himself through cognitive behavioural therapy when he found himself staring at logging trucks, compulsively calculating what was on the tray and where it was heading. 'I was going crazy with fury.'

But support for Manning and his lonely stance against the industry has come from unexpected places. He recalls receiving a congratulatory phone call after appearing at the Senate inquiry from a logging boss in Smithton. Manning had supervised his logging operations many years previously and even closed them down once because of poor practices. Their working relationship had been fraught with yelling and tension. But this phone call was different. 'It was the best work you've ever done,' the logging boss said of Manning's evidence at the enquiry.

Before I leave, Manning brings an odd collection of plaques up from his basement to show me. Lumps of lacquered wood, they are awards for his service to forestry. There is one for twenty years' service, another for thirty. He puts one on his shoulder and poses for me.

'What do you think?' he says. 'Reckon I've got a chip on my shoulder?'

Laughing, I nod. 'Maybe. You're pretty stubborn.'

He smiles. 'Yeah. I like that about me. I won't let go.'

*

When I ask Barry Chipman about the policy of self-regulation in Tasmania's forests, he scoffs angrily, snapping his head like a horse with flies.

'It is ridiculous, this accusation of no regulation. We have forest practices officers on each coupe who report back to the Forest Practices Authority any environmental issues they believe will arise from logging the area, as well as a detailed plan to work with.'

I point out that it is now well known that timber companies such as Gunns employ these officers.

'Many people have two jobs,' he shoots back.

I suggest that even the intermittent presence of a forest practices officer won't necessarily inspire a sense of responsibility in contractors if their agreements are overly focused on extracting as much pulp as quickly as possible. If I were strapped for cash and paid per tonne, I doubt I'd stop mid-coupe and say to my workmates, 'Gee, I think those are some rare trees' or 'Look up there – is that a nest? We should stop and call an expert.'

Chipman snorts and shoots me a look as if I'm immoral for even thinking this. 'That would require me to question the integrity of my colleagues. We have what's called Internal Peer Review. It's an issue of trust.'

When Channel 9's *Sunday* program covered Bill Manning's dismissal, the reporter, Graham Davis, asked Graham Wilkinson why Manning was directed to withdraw his notice. Wilkinson replied that Manning had been undermining his fellow officers. If he had concerns, Wilkinson said, his job was to put them in the audit. 'That report would go to the board and it would be published in our annual report, as they were.'

When Davis queried the usefulness of a random audit of 15 per cent of forestry plans, when rare and endangered species might already have been cleared, an almost Orwellian exchange occurred.

'Well, that comes down to the knowledge we have of where these species are,' explained Wilkinson. 'We have very good databases.'

Flabbergasted, Davis said, 'After this, they're gone. There's nothing there!'

'Well, they're still on a database, Graham,' Wilkinson replied.

Davis also brought up the case of another coupe, at Reedy Marsh, where Gunns was planning to convert native forest to plantation. Of seven plant communities on the site, the timber company's forestry officers mentioned only three in their logging plan, omitting a rare variety of eucalyptus that was protected by a moratorium. Suspiciously, the presence of the rare plants meant the site was especially fertile – perfect for a plantation. Only when a local activist entered the site, documented the species and raised a huge public fuss was the logging plan withdrawn. Similarly, it took Davis's news crew to get another coupe re-inspected, after Davis mentioned in his report they had come across a breach. The Forest Practices Authority invited him to bring the footage in; they would have a look and decide if the contractor ought to be fined.

Gemma Tillack of the Wilderness Society is also exasperated by the FPA. She says it is members of the community, activists and environment organisations who end up monitoring the coupes, rather than the authority's staff. 'I've been into a logging coupe, after fighting to get a hold of the forest practices plan, and then taken photos, evidence, measurements, the gradient of the slope they logged on and so on, to show they've acted outside the plan. But that is a role they should be doing.'

Pete Godfrey, an electrician from just outside of Deloraine, works three days a week and spends the rest of his time independently documenting the clear-felling and aerial spraying in his area. For the last five years he has regularly gone out to document and report breaches of the Forest Practices Code, taking photos of landslides and swirls of silt and debris pushed into streams.

'I've managed to get the Forest Practices Authority to fine Gunns and Rayonier on one occasion each, and to make Gunns do some pretty extensive remedial work.' In turn he too is treated as a pest by the FPA.

In the north-east, one resident noted that while their community blockade had failed to protect one particular forest, their actions and consequent local court appearances ensured work in the coupe became much more careful: 'They knew we were watching them.' But residents also report being nudged off the road by trucks, even with their children in the backseat. 'I peeled every sticker off the rear window after that,' says one mother from St Marys.

When the police watched the activists' footage of the sledge-hammer attack on the car with Miranda and Nish inside it, they paused on the image of one figure in particular. It was the Upper Florentine's forest practices officer, an employee of Forestry Tasmania. The state employee had denied being in the forest when the loggers circled the car and smashed it, but when the footage is slowed down, he shows up clearly. The same forest practices officer is now up on public nuisance charges after he claimed the activists had assaulted him in a separate incident. Again, footage showed otherwise, revealing he had driven his car across their path and cornered them against a fence, albeit a fence one protester was chaining himself to.

But despite the threats and the agro, more people keep emerging to defend the forests from the chippers.

THE BLUE-CARROT STATE

While I am meeting with Gemma Tillack at the Wilderness Society's Hobart office, an old man knocks on their back door. His hands are trembling.

'They're killing all my animals,' he says to one of the head campaigners. 'I don't know what to do. They're killing them all.' Wringing his hands, he stands on the cement drive, next to the group's urban vegetable patch. His pale blue eyes are rheumy and the skin around them crinkles like dough. 'I'm building a library out there. I've a shipping container of books. At night I take walks through the bush and see all the creatures.'

Gemma Tillack tells him to come in and makes him a cup of tea. Tom tells us he was brought up in the hydro camps and grew up near the dolerite cliffs of the west coast.

'I used to be able to fight,' he says, stiffly bending his fingers into a fist. 'I even made some money from it. But I'm too old now.' He had moved inland to retire. His pocket of land was once surrounded by forest. There had been no real need for fences – they were just flimsy gestures. But then, plantations cropped up around him and the creatures flooded into his small pocket of land, spatting for territory.

Soon, however, they were lured back by the rows of new saplings. When piles of corn and vegetables were dropped around the plantation, Tom knew what was coming. The shooters arrived.

'The first two times they warned me,' he said, recalling the staccato of bullets. 'But the third time, the fella didn't bother telling me. I was out on my walk when the shooting started up and I had to yell out that I was there.' Shooters are employed to stop animals nibbling on the new trees. They zoom around on quad bikes with spotlights. 'I spent a lot of money trying to bring them back with my own piles of corn and vegetables. Why don't they just build better fences?'

Tom would survey the damage the morning after each shooting, checking the dead animals' pouches for babies. 'They have to count their cull, but I doubt they include the pouched young, especially the ones in the sealed pouches. It's heartbreaking. Most of the time the babies are already dead. But still I take them out of their pouches – they're the most beautiful things. Seems the only time you get really close to the animals is when they're dead.'

Tillack takes notes as Tom talks, closely questioning how many times he was called before the shooting began, how much warning they gave him and whether he had any say in the shooting times.

'Oh no,' he says, 'The first two give me a day or two warning but didn't say when for certain. I had to phone a couple of my friends and say don't pop in the next few evenings. The third guy didn't bother telling me at all.'

This is not the first time Tom has watched a place emptied of life. 'I used to visit this little bit of forest out west, camp there and stuff. I loved that place. There were wombats, pademelons and masked owls everywhere.' When he returned one month, he found the area flattened with little pink 1080 poison signs picketed into the ground. He looks at me, voice wavering. 'It's the most horrible thing. Seeing all those animals dead. Especially when you've gotten used to walking around at night and watching them move around and squabble and have their young. It's awful.'

*

Terry Rousell remembers the first month Forestry laid down 1080 in Ben Nevis.

'It was in a coupe that used to be a 2000 to 3000-acre cattle run. It was a big basin of grass, and it was the eeriest thing to walk into – it was like entering a morgue. There were dead animals everywhere – deer, wallabies, wombats, possums, kangaroos, rabbits, birds. Some hadn't died yet, and were frothing at the mouth, bumping into things. After the time we found a bunch of dead wallabies in the water catchment up from our drinking water, I sent a letter to Bryan Green, our Labor MP, and he wrote back saying there was no need to worry, that 1080 doesn't harm humans.' Rousell looks at me incredulously. 'I mean, where do these politicians come from?'

People soon started to find listless wallabies in their yard, falling over, trying to stand back up, eyes mucus-coated and glazed. Forestry issued muzzles after pet dogs were found dead. Also among the casualties were endangered species such as bettongs, quolls and potaroos – small brown marsupials that move like rockets.

After the first few years of laying down 1080, Rousell says, the forestry industry smartened up, for appearance's sake. 'They lightened the dose so animals weren't dying in streams and were able to get back to their burrows to die quietly.' The poison inspired another bumper sticker: 'The Blue-Carrot State,' referring to the practice of baiting wildlife with carrots soaked in 1080, which gives the vegetables a blue, bruised tinge. A Hobart medico, Michael Vaughan, was famously denied a custom numberplate – NO1080 –

for his ute; the motor registration authority said it was too political. Today, Forestry Tasmania isn't allowed to lay 1080 on state land, but it is still used on private land. Gunns announced plans to stop using 1080 in its plantations only in June 2010.

Barry Chipman's view is that forestry should be required to avoid direct harm, but no more; demands that the industry go out of its way to protect wildlife are commercially paralysing. 'We now have this idea that you need to *protect and enhance*,' he told me. 'Any activities, by anybody, need to protect and enhance the survival of the species, and that's unrealistic.'

According to a report by the Tasmania Conservation Trust, over 200,000 wild animals including wallabies, quolls, potoroos and wombats were killed between early 2002 and April 2004 by 1080. But as Rousell's theory suggests, that is only the number of dead animals *found*. When the Co-operative Research Centre for Forestry undertook a 'poison lethality' study, using collar-tracking devices to monitor native animals, they found that the death rate could be much higher. The collars demonstrated that 75 per cent of the animals managed to hide themselves after consuming the poison, their carcasses ending up in dens, burrows, bales of hay, hollow tree trunks, and even buried beneath fallen leaves.

Androo Kelly, of the Trowunna Wildlife Park, believes the use of poison baits should be stopped altogether.

'I guess my own views are purely anecdotal,' he says, 'but it comes from a constant source. This place.' He gestures at the park in Mole Creek. 'The biggest change I've seen out here is the huge reduction in small birds. Aggression has been rewarded and the more vulnerable creatures are disappearing. Immunology seems to have weakened and the animals don't heal as well. Even the tough birds such as the peregrine eagle have their share of problems.

After they eat carcasses killed by 1080, the poison attacks their calcium stores and their egg shells are too weak.'

Alison Bleaney, a doctor in St Helens, recalls seeing a frothing wombat, followed by her cub, bumping into shrubs and a farm fence.

*

Everywhere I go I meet locals intricately connected with the island's wildlife. Some people have creatures literally hanging off them. With others it is more subtle: only well into our conversation do I notice the tiny unfurling of a sugar glider in the fold of a shirt, or the joey in a pillowcase. At night some Tasmanians speed along the dark roads, knocking off animals like insects, while others drive slowly down the same roads each morning, stopping to check the pouches for babies and to drag the kill off the roads so the devils can eat in safety. Visitors to Tasmania often comment on the amount of roadkill they see. They dwell on it, express their horror at it, before returning to their empty suburban streets with their gangs of crows and noisy miner birds.

Almost every local turned environmentalist I meet has had to kill an animal in his or her life. Vica Bailey of the Wilderness Society remembers, as a kid on the family farm, tagging along with the men as they rounded up a mob of forester kangaroos and herded them into a line of gunfire.

When he got older, he went out shooting with his mates. But then something changed.

'I'd been away for a year and had just gotten home. A close mate of mine had died while I was overseas and I felt bad that I hadn't been here.' Inevitably, a mate asked him to go out shooting. 'It was what we used to do, so I never really thought that I couldn't do it anymore.' He lined up a wallaby in the spotlight and shot it dead. 'It made a sound before falling over. I remember looking at my mate and saying, "I can't do this."' His friend turned the ute around and they headed home.

Another local tells me he used to go out shooting, until one day his shot was off and he hit a rabbit without killing it. 'It just started shrieking. I didn't even know rabbits could make a sound, but this one was just screaming.' He tried to put it out of its misery, but it got away, shrieking the whole time. 'I couldn't shoot anything after that,' he says.

But these are private experiences, quiet moments, born of solitude and contemplation. Most shooting down here takes place in groups – there is no space or time to hear the death of an animal, to study its eyes as life burns out like a match – there is just a lot of cheering, whooping and tooting of horns. Now that the land has been leased by timber companies, Terry Rousell tells me, private 'shooting parties' are often held in the north-east.

A local, Lyn Hayward, who with her husband has turned their property just outside Deloraine into an animal refuge, sums up the irony of these two types of people living side by side. 'As one wildlife officer said to us, in the same evening he could drop off an injured animal to a carer and then half an hour later, give out a 1080 permit.'

*

In 1972, Christopher Stone, a law professor at the University of Southern California, wrote an article titled 'Should Trees Have Standing?' In floating the radical concept of extending rights to nature, he observed that the extension of rights to beings considered inferior and incapable invariably seems absurd before it happens. Female suffrage and the emancipation of black Americans were once unthinkable. Australian Aborigines had been granted the right to vote in federal elections only ten years before Stone published his essay, and even then it took decades for the first Australians to be recognised as equals by the majority of white society. In Tasmania, homosexuality was decriminalised in 1996 despite cries that the change would encourage 'moral pollution,' or that decriminalisation was pointless because there were no gays in Tasmania.

All these rights, however, were already enjoyed by humans in some shape or form at the time of reform; reform involved extending them to a greater part of the population. Even when rights are granted to animals, it is because we believe they suffer pain as we do or can count up to ten bananas and therefore have a smidgeon of 'human' intelligence. It's quite a leap to extend the concept of rights to trees or the wilderness generally – perhaps especially so in a place like Tasmania, where the 'taming' or 'conquering' of nature plays such a crucial role in the colonial story. Michael Field, the former Labor premier, tells me it was tin mining on the island's west coast in the 1890s that finally freed Tasmanians from their 'convict stain.' In the ramshackle mining towns, he says, power was wrested from the gentry and the landowners and returned to the 'people' – a.k.a. the Labor Party. I start to wonder if the defensive and aggressive pride I keep running into stems from that time. Is this 'I built this place and it owes me' attitude the culmination of a long struggle for equal access to the island's resources? Did equal rights for all white Tasmanians mean equal opportunity to own and exploit the environment?

For environmentalists, this sense of 'owning' nature is difficult to understand. Nevertheless, the Greens are increasingly aware of the need to frame environmental arguments in terms of human economic consequences. The ocean is a 'resource' to be protected to sustain the fishery industry, the forests are important 'carbon sinks,' and so on. But such arguments have the potential to backfire: in the hope of saving old-growth forests, the Greens spoke of 'value-adding' and 'job-creation' with plantations. As a result, the state now has vastly more plantations in its water catchments and on previously forested land, not to mention the proposed Bell Bay pulp mill.

Hence Stone's argument that nature should be accorded its own rights. Stone is quick to clarify that this wouldn't mean humans could no longer chop down a tree or catch a fish, or that courts would be clogged up with cats suing dogs for harassment. That would be as silly and absurd as letting an infant vote because it has 'rights.' An infant does, however, have a right to exist, unto itself and for no one else. Similarly, nature might be accorded a right to exist simply for its own sake, rather than only when and where it meets the needs of humans.

In 2009 in Shapleigh, a small American town in Maine, residents voted to grant the town's natural assets legal rights. Like other towns nearby, Shapleigh is trying to protect its aquifers from the Nestlé Corporation, which relies heavily on the region's springs for its bottled water. So far, no town had managed to protect its water. As with the Wilderness Society, there were internal power struggles among the head campaigners; as with Timber Communities Australia, there were false community groups; and as in Tasmania, there were endless protest actions, violence and division, while trucks loaded up with the source of the turmoil continue to roll on by. In Shapleigh, however, any resident can now 'stand' in court to seek damages on behalf of nature. Other towns in the US have enacted similar statutes. In Europe, too, a campaign is afoot for a charter of 'planetary rights' to sit alongside human rights at the

United Nations. Two years ago in Ecuador, a new national constitution was drafted recognising nature's 'right to exist, persist, maintain itself and regenerate its own vital cycles, structure, functions and its evolutionary processes.'

There are approximately twenty-two significant wildlife diseases currently wreaking havoc on Tasmania's wildlife. Some are mysterious such as the devil facial tumour disease, which has killed 70 per cent of the devil population since 1996, while others are well known and ordinarily insignificant. But vets, park operators and wildlife carers note that even in the case of these small infections, animals and birds don't seem to be healing. Wombats are under attack from mange and platypuses covered in a fungal disease.

One local veterinary pathologist, Dave Obendorf, has compared the mass deaths of devils, oysters, even sparrows, to that of canaries in mine shafts. 'They're telling us something and we're refusing to listen,' he tells me. In St Helens, a town on the north-east coast, Alison Bleaney, the local GP, says she has noticed an odd blip in human cancers, unusual and rare ones at that, and not within their usual demographics.

Dr Bleaney believes more transparency is needed on the island when it comes to potentially harmful activities such as converting forests to monoculture and chemically dependent plantations of 'genetically improved' trees. 'We need to know who's doing what, not just to join the dots but to eliminate causes as well.'

'We've already lost one carnivore and we're on our way to losing the next two, the devils and the quolls,' she says. 'And if you compare the devils' life expectancy to our own and they're going boom boom boom' – she clicks her fingers to indicate the species dying off – 'it could take a generation or even three to tell for certain if and how we could be affected.'

THE SHAREHOLDERS

At the beginning of the twenty-first century, the Wilderness Society decided to change its tactics. They would get tough, corporate-style. With the help of leaked documents and tip-offs, they began to focus on shareholder activism and raising customer awareness. They lobbied Gunns' shareholders, customers and buyers, institutional investors and potential financiers. They urged people to contact their super funds and make sure none of their money was going to Gunns, directly or indirectly. As word spread, customers across Australia began contacting their bank managers.

To begin with, the corporate action seemed like using slingshots and stones against the company's great metal hull, but with help from unexpected places the Wilderness Society soon started making dints. An anonymous international business strategist, reported to be employed by some of Australia's biggest companies, gave advice on building awareness in Japan. Fund managers helped draft resolutions for AGMs and lawyers provided pro bono advice. The Society even claimed to have a 'corporate raider' on board, a shareholder who infiltrates and disrupts a company. In the 1980s, these raiders were romanticised villains, in it for the money and the thrill of it – think Michael Douglas in *Wall Street* and Richard Gere in *Pretty Woman* – but today they are more likely to be activist shareholders, motivated by ethical concerns. Leanne Minshull, formerly one of the Wilderness Society's head campaigners and one of the Gunns 20, would not name names when asked who was advising the Society. 'Everybody is paranoid and I don't want to piss this up against a wall,' she told the *Age*'s Leon Gettler in 2003. John Sevior, head of Australian equities at Perpetual Trustees, met with the Society; he told Gettler it was the first time he had seen such a campaign and suggested there could be more to come. 'The world is getting more determined in a lot of ways,' he observed.

In 2003, using new corporation laws, 100 activist shareholders, controlling about 0.3 per cent of Gunns' entire stock, called an extraordinary general meeting. John Gay, who at the time owned 5 per cent of the shares, denounced the meeting as 'invalid,' but the company was soon forced to concede. Before the meeting, the activist shareholders raised a motion that Gunns withdraw from 240,000 hectares of high-conservation-value Tasmanian forest. A date was set and key shareholders were lobbied. First in the campaigners' sights was the Commonwealth Bank, which at the time held a 17 per cent stake in Gunns. The campaign targeted individual branches and account holders, as well as investors nationally and internationally. The Wilderness Society took investors on daytrips through forests earmarked for logging. Customers began to move their accounts and to enquire about their super funds.

The campaigners needed 75 per cent of the vote to pass a resolution at the shareholders' meeting, and knew this was unlikely – at least for now. On the day of the meeting in Launceston, log trucks lined up around the block, bellowing their horns, while activists crowded the sidewalk. The press and general public were banned from the meeting, but Stephen Mayne, a shareholder and writer for *Crikey*, was present and wrote about it afterwards. The attendance was too large for the company boardroom, so 400 or so shareholders and proxies were herded into a huge kiln at the back of the plant. The directors refused to provide microphones, chairs, coffee or tea, or to feign even remote interest in the meeting; Gay mumbled for a few minutes before trying to launch straight into a vote. When Mayne asked if Gunns could explain the financial impact of stopping logging in eight specific forests, Gay 'claimed Gunns hadn't run the numbers and would only ever do so if it became a reality.' Mayne later wrote that it was this lack of financial analysis that prompted shareholders BT Financial Services (owned by Westpac), Unisuper and Local Authorities Superannuation to abstain from the vote.

It is generally accepted that significant shareholders are unlikely to vote against the economic bottom line. For these shareholders, abstaining is as radical an action as they're likely to take, often citing insufficient evidence. It is a cautionary stance taken by a shareholder unwilling to lose money but nervous about their reputation. Abstentions were what the campaign was aiming for, but the result wasn't encouraging. The number of votes withheld represented only 2 per cent of the total shares, with major institutional shareholders such as Perpetual Trustees, Commonwealth Bank, Deutsche Bank and AMP coming out in support of Gunns' current practices. It was now clear which companies were in favour of logging Tasmania's forests; the corporate activists had their work cut out for them.

Within two years of the extraordinary general meeting, the Commonwealth Bank sold almost 400,000 Gunns shares. Reports prepared by CommSec, Commonwealth Bank's share-broking service, advised shareholders to reduce their holdings in Gunns, concluding that the pulp mill was a high-risk investment. According to

CommSec, the project could only be viable if world pulp prices remain high. Although Chinese demand for pulp will surge in the next decade, Gunns will be competing against South American pulp mills, which can produce pulp at almost half the price. In June 2010, the Commonwealth Bank bailed out altogether.

The chairman and executives of ANZ – not a shareholder, but one of Gunns' key financiers – agreed to meet with the campaigners and visit the island's forests. ANZ later pulled out of negotiations to finance the pulp mill. To date, more than fifteen leading banks once rumoured to be in discussion with Gunns, including Deutsche Bank, the Royal Bank of Scotland and the Bank of China, have publicly declined to finance the proposed mill.

Meanwhile, an awareness campaign was launched in Japan, with Greenpeace and the Wilderness Society teaming up with Asian green organisations. Woodchip buyers and papermakers, including Nippon Pulp & Paper, Oji Paper and Daio, as well as consumers, were asked to consider where their product came from. Feeling besieged, the president of Nippon, Takahiko Miyoshi, wrote to Tasmania's premier at the time, Paul Lennon, urging him to fix the divisive forestry debate. 'Never before have issues that are essentially local Tasmanian issues come to interfere with our business and cause confusion with our own paper customers here in Japan,' he wrote. Scandal befell Nippon and Oji after revelations that they had been misleading their customers about the recycled content in their paper, falsifying information for over ten years – in one case using only 1 per cent recycled paper in a product which claimed to be 40 per cent recycled.

Back home, the campaign got a national platform in 2007 when Geoffrey Cousins, a prominent businessman and former prime-ministerial adviser, took up the activists' cause. After reading

Richard Flanagan's *Monthly* essay 'Gunns: Out of Control,' Cousins was moved to letterbox 55,000 copies of the essay in the electorate of Malcolm Turnbull, then the environment minister responsible for approving the proposed pulp mill. With his boardroom prowess, Cousins became involved in organising meetings between major Gunns shareholders and the Wilderness Society. Behind the scenes, he began to push hard for the resignations of John Gay and Robin Gray.

*

Just two weeks before Jim Bacon died, Peter Cundall – long-time host of ABC's *Gardening Australia* and anti-Gunns pulp mill campaigner – says, he received a phone call from the former premier.

'It was in no way hostile. Just the opposite. It was two weeks before he died and his voice was already quite weak. He implored me to keep fighting against those who were "destroying and poisoning Tasmania."' 'Keep up the fight against the pulp mill, Peter,' said the instantly recognisable voice down the phoneline. 'I can't stop them. They run everything.'

When I hear this story, I am stunned. But most Tasmanians I meet, including Cundall, seem unperturbed by it.

'Oh yeah,' says one Greens campaigner when I ask if he's heard about the conversation, 'everyone knows about that' – as if there is nothing odd about a dying ex-premier phoning up a celebrity gardener and telling him the state government is a sham.

Many suggest Bacon might have been seeking redemption. He and Cundall were both 'comrades' back in the day, but Cundall stayed true to his belief in social justice while Bacon strayed, seduced by the scent of power. When I first ask Cundall about the call, he is grumpy that he even has to repeat the conversation.

'Like I've said many times before, it was an amicable conversation, nothing sinister about it. And when I politely pointed out to him that when he was premier he had virtually let the destructive, greed-driven woodchip and logging industry loose in our beautiful native forests, Bacon responded by saying: "I was the premier of Tasmania but these bastards were infinitely more powerful than me. You've no idea how powerful they are. I couldn't move. For God's sake, keep fighting them. That's why I'm ringing you – they have to be stopped."'

THE MILL

THE PROPOSAL

I n a coffee shop, I meet a quiet, conservative scientist. He is neither on the side of Forestry nor the environmentalists. He is a perfect picture of objectivity. Over cups of tea, with rattling activity around us, he implores me to get the science right and then leans back enigmatically.

'There is one way,' he says, 'of solving the forest conflict on this island.' I sit forward excitedly. 'The pulp mill,' he continues. 'The mill will put an end to this debate once and for all.'

A forest management scientist with the CSIRO, Dr Chris Beadle's eye has been caught by the proposed mill's gargantuan appetite for woodchips: 4.5 million tonnes per year. After examining documents submitted as part of the mill's assessment process, he concluded that, contrary to Gunns' claims, it just wasn't possible for a project of this size to be primarily plantation-fed.

'If it goes ahead, this pulp mill as it stands will put an end to the forest conflicts once and for all,' he repeats. 'Because there will be no wood left over. No further timber negotiations, value-adding or innovation will be possible. This mill's appetite will need all that is available.'

*

To begin, there were two locations in mind for the new mill. There was the Bell Bay site in the Tamar Valley, a wine-making and tourism region in the north-east with a population of 100,000 people. The second site was Hampshire, a north-western district just behind Burnie. Comparatively isolated, Hampshire has very few residents and is surrounded by plantations – an acceptable location for a pulp mill, according to the scientist and mill expert Dr Warwick Raverty. The Hampshire site was, however, abandoned by Gunns, on the grounds that transportation costs would make the venture unviable.

It was cost-cutting measures such as this that had financial analysts wondering about Gunns' ability to be competitive in such a high-volume, low-value market. Naomi Edwards concluded that the company's set-up and ongoing costs would be significantly higher than those of emerging pulp producers in South America, where Australia's native trees are said to grow faster. In such a cut-throat commodities market, it was therefore no wonder that the timber giant made the Tamar Valley its preferred site. But it avoided one costly disaster only to crash headlong into another. People. For while opposition to the pulp mill mightn't have the same romantic appeal as protection of the Franklin River or Lake Pedder, its ramifications threaten to go far beyond the 'green' movement.

On the eastern shore of the Tamar River, 36 kilometres from the city of Launceston, the pulp mill site at Bell Bay sits alongside an existing Gunns woodchip mill. There are two ways to see this coastal valley on the northern tip of the island. One is as a sweeping vista of vineyards, farms and forested hills with restaurants and wineries tucked into levelled nooks. There are postcard moments: lonesome sheds that look like they've been built out of driftwood, with vast views of the river as it snakes through the wetlands and out to the Bass Strait, where fishermen bring in their catch. This perspective has given rise to a broad coalition of convenience against Gunns, with grape-growers, fisheries, residents, restaurant owners, hoteliers, bed & breakfast owners and walnut farmers lining up with forest activists, the Wilderness Society and the Greens.

Supporters of the pulp mill, by contrast, see the proposed location as a heavy industrial precinct. They say the mill's opponents are promoting a falsely pure view of the region, and that the pulp mill will not be the only smudge on the horizon. In George Town, the district closest to the pulp mill site, you can drag your fingers along sticky black windowsills. Industry is a part of life here: there is already a deep-water port, an aluminium smelter, a seafood processing facility, sewage treatment plant, ferro-alloy processing plant, sawmill, woodchip mill and power station. George Town has high unemployment, but instead of being cynical about the promises industry has made to the district over the decades, residents are optimistic about the increased capital and employment that construction of the mill might provide.

In an ode to the mill, Paul Lennon said it would deliver Tasmanians from the 'bondage of poverty,' and repeated Gunns' prediction that it would create 2,000 temporary jobs during construction, 1,617 indirect jobs and 280 jobs in-house.

'Two thousand Tasmanian families,' the premier declaimed, 'would not have to worry about how they are going to pay the bills or afford to send their children to school or wonder where their next meal is coming from. Two thousand Tasmanians who will not have to leave the state to find work. The benefits are also within our reach. The opportunity is there for us, but we cannot afford to let it slip.'

The mill would mean an extra $870 each year for every Tasmanian household, he added.

In the same year – 2004 – that Gunns announced the mill, British travel agents ranked Tasmania as one of the world's top four hotspots to visit. Over one million visitors spend more than a billion dollars on the island each year. And in comparison to the elusive 6000-odd jobs in the island's timber industry, more than 30,000 Tasmanians are employed in tourism. Saul Eslake, born in Tasmania and formerly the chief economist at ANZ Bank, pointed out that Tasmania's recent feat of achieving the lowest unemployment figures in the country had little to do with the mining, forestry, fishing and agricultural sectors, which contributed less than 5 per cent of the new jobs in the past five years. According to Eslake, employment is moving away from commodities and towards niche-based businesses and industries, relying on strong customer service and quality. 'That doesn't mean that commodity-based industries have no part in Tasmania's future,' he comforted. 'It's just that it's unlikely to be a significant one.'

But state leaders seem constantly to default to infrastructure spending to create a mass of short-term jobs. Tom Ellison, general manager of the advisory firm Wills Financial Group, recently noted the problem with this approach. 'Building roads and bridges with Commonwealth funding might provide a short-term economic benefit for the state,' he warned. 'But infrastructure spending is a one-off; once completed, the jobs dry up.' Like Naomi Edwards, Tim Harcourt, the chief economist at the Australian Trade Commission, expressed concern that the state government was tying its fate to a cheap and high-volume exporter of mostly unprocessed commodities in a market where competitors have vast advantages. But all of these economists, actuaries and analysts were ignored.

*

In 2004, when John Gay announced the pulp mill, a short time after sending out the Gunns 20 writs, he said it would be primarily plantation-based. At start-up the majority of the wood would come from native forests, he conceded, but the switch to plantation pulp would come within ten years. 'Only world's best technology utilising a low-impact "totally chlorine-free" (TCF) mill will be looked at,' the timber company stated. But then, fewer than six months later, on the last day for public comment and a quarter of the way into the project's assessment process, there was a change of plans. The TCF pulp mill was now to be 'elemental chlorine-free' (ECF).

Why did Gunns make the switch from a totally chlorine-free mill to elemental chlorine-free? The answer is in the trees. Trees are generally made up of cellulose fibres bonded together with lignin. In paper and pulp making, fibre is the sought-after ingredient. Lignin is a glue-like substance that makes trees strong and adds to the durability of consequent building products, but it is dark – hence the colour of wood. Manufacturers of brown paper bags and cardboard boxes aren't troubled by lignin, but makers of paper – for whom Gunns' type of pulp is intended – want their pulp 'whiter than white.'

If the mill's wood were sourced from plantations, a TCF process could be economically feasible. But native-forest timber, especially mature trees, has very high levels of lignin. Using the TCF method on it would be expensive, difficult and potentially produce a lower quality pulp.

It began to dawn on Tasmanians that Gunns had no intention of changing its approach. Not only was there the shift to the ECF process, which implied a native-forest-fed mill, but it was also proposing to build a wood-fired electricity generator alongside the pulp mill to generate power and direct left-over electricity back into the grid as 'renewable energy.' Journalists and forest activists picked over the company's forecasts and wood-supply agreements to discover that not only would the timber giant continue its annual export of 3.4 million tonnes of woodchips and consign half a million tonnes of timber to the electricity generator, but it was planning to push a further 4 million tonnes through the mill. At this rate the project looked likely to double the island's current rate of clear-felling – and Gunns would need to access state forests for at least another thirty years. Lennon's claim that Tasmania needed the pulp mill to 'stop this lunacy of sending our woodchips offshore' was beginning to look like a ruse.

In desperation, Chris Beadle wrote to the *Australian*. 'There does not appear to be enough wood to meet what the proponent is saying will be sourced from plantations. If I'm right, there will be further pressure on native forests over and above that which they say.' Speaking to the media is most certainly not Beadle's preferred means of communication, but when he broached the subject with the timber industry's spokespeople and the government's Pulp Mill Taskforce, he was scoffed at. 'I was treated with derision. I'd done the maths and it just isn't logically possible to feed a mill of that size with the island's current and future timber resource. But when I presented it to them, they told me not to be stupid and of course they weren't going to pay for wood to be imported to the mill.'

Concerns about the mill's appetite were accompanied by growing environmental alarm. Local businesses and residents, some of them fewer than six kilometres from the proposed location, feared that the projected daily emission of '300 kilograms particulate air pollution, which will include non-condensable gases' could translate – as it has done in many other pulp mills around the world – into a mixed aroma of rotten eggs, boiled cabbage and burning rubber. The region has almost half of the island's wine industry, and wineries grew even more worried after discovering that ECF pulp mills can emit chlorophenol compounds, which are known to contaminate the taste of wine, giving it the bitterness of disinfectant.

The island's fishing industry, which generates approximately $400 million each year, was equally concerned about marine effluent. The ECF process is said to be '*virtually* dioxin-free,' but it is not entirely free of the risk of producing these carcinogenic substances. The mouth of the Tamar River, with its unique whirlpools, estuaries and the shallow Bass Strait is, to date, an untested environment for adequate dispersal. It is this particular issue that repeatedly prevented Gunns getting full approval from the federal government, despite its reassurances that treated effluent from the mill, a maximum of 64,000 tonnes per year, will be 'discharged through a multi-port diffuser system at a depth of 26 metres, about 3 kilometres into Bass Strait in an area with minimal aquatic activity.'

Other concerns included the high increase in numbers of log trucks on the roads and the mill's proposed massive water intake, which was nearly double the combined usage of Launceston, Georgetown, the West Tamar and Meander Valley councils, including industrial customers. Finally, the Australian Medical Association was worried about the mill's contribution to Launceston's already notorious inversion layer, a natural phenomenon that traps smoke from industry and home wood fires.

Everything was in place for a pitched civic battle. All that was needed was a central forum. The proposal was now handed to Tasmania's independent planning authority, the Resource Planning

Development Commission (RPDC), to be assessed by an expert panel. Members of the panel included Julian Green, a senior bureaucrat, and Warwick Raverty, a former VISY paper consultant known as the country's 'go-to' pulp-mill expert.

THE PANEL

I t was late 2004 and the fate of the mill was in the hands of the RPDC. Well, it would have been if Premier Lennon could let go of it. In 2003, before Gunns had even announced its plans for the mill, Labor had launched the $1.6 million dollar taxpayer-funded Pulp Mill Taskforce to convince the public that the island needed such a project. The taskforce slipped into the absurd when it unveiled a 'Pulp Mill Information Bus' and proceeded to have it driven around the state.

Lennon thought nothing of appearing on pro-pulp-mill TV advertisements. Even more peculiar was his decision to continue with the taskforce once it became public knowledge that Gunns was behind the proposed mill. But the taskforce was not the only source of friction for the RPDC; there was also the mill's hostile and unco-operative proponent, Gunns. Within a year, John Gay was complaining of costly delays caused by the assessment process, even though it had been agreed at the outset that the process would take at least two years, and within six months, a surprise announcement from Gunns that the mill would no longer be totally chlorine-free forced the panel to drastically revise its guidelines and timeline.

Then in December 2006 Dr Raverty resigned from the panel. The Labor minister Steve Kons announced that as the forestry scientist had previously advised Gunns on a 'separate issue,' he had been advised to resign to avoid the perception of bias. Less than a month later, the head of the RPDC, Julian Green, resigned too. Both ex-panel members cited political interference and bullying as reasons for their departure. A senior public servant revealed to the *Mercury* that Bob Gordon of the Pulp Mill Taskforce had warned Julian Green that any complaints he had about the taskforce would fall on deaf ears, since Lennon was a close mate. 'Go ahead,' Gordon allegedly said of the premier. 'He's my mate, you know, he won't do anything.'

For his part, Warwick Raverty alleged that Les Baker, a senior executive at Gunns, had accosted him at the Canberra airport and only stopped yelling and swearing when Raverty threatened to get the police. Baker's last words, said Raverty, were along the lines of 'You've lost all your friends – they're all laughing at you! You are going to lose this fight and you are going to be a loser for the rest of your life.' Of his time spent on the panel, Dr Raverty said the timber company's impact statements were repeatedly riddled with inaccuracies. 'Rather than the succinct information that could be easily digested, what Gunns published after a delay of three months was, in my opinion, a 1,200-page dog's breakfast of poorly presented data and statutory declarations.' At one point the timber company bizarrely predicted air pollution in the Tamar Valley would be lower *with* the pulp mill. 'I've come to the sad conclusion that Gunns is not a fit and proper company to build a pulp mill anywhere,' he stated.

In the midst of all this, the state was busily assigning replacements to the RPDC and its panel. Simon Cooper, a lawyer with strong family Labor connections, was appointed interim head of the commission, while Christopher Wright, a retired Supreme

Court judge, was to chair the assessment panel. Around the same time, a two-year worldwide search for a new director of Forestry Tasmania finally ended. Bob Gordon, the head of the Pulp Mill Taskforce, got the job. It wasn't long before Wright accused the premier of 'completely inappropriate' attempts to pressure the independent assessment. Wright claims he was told in a meeting with Lennon and Gay that the process was to be finished by July 2007, not November as previously hoped, in order to meet Gunns' commercial timetable. Wright says he was also told by Lennon to 'water down' the inquiry and claims he was given an ultimatum to drop public hearings and wind up the assessment, or face the panel being dumped in favour of fast-tracking legislation. In March 2007, Wright called a press conference. The former judge released a statutory declaration to back up his version of events and his letter of resignation.

Lennon denied making such demands. 'I hope Christopher Wright now explains why he could not shave any time off because that will be illuminating, quite frankly, now that he has entered the public debate,' the premier told the parliament. The acting RPDC head, Simon Cooper, later revealed that the premier's secretary, Linda Hornsey, had invited him to publicly contradict Wright's claims. 'I wasn't going to hop into a credibility contest with Mr Wright. In any event, I happened to agree with him,' said Cooper.

By the end of March 2007, the RPDC assessment process was over. Gunns withdrew, saying the delays were 'commercially unacceptable.' Gay said the company had already spent $11 million (albeit a burden eased by a $5 million gift from the Howard government) in putting together almost 10,000 pages outlining the mill's impacts and was no closer to getting a decision. Critics, however, suspect the timber company was tipped off so it could pull out of the RPDC process before it failed the assessment. A few days earlier, Cooper had phoned Linda Hornsey to say the panel had found Gunns 'critically non-compliant' and that the commission was going to inform them of this. He was told to hold off from making the call until she got back to him. Hornsey later said that she had believed she could bring about a more positive reaction from Gunns, but Cooper believes 'that holding off on sending the letter allowed Gunns to withdraw from the process without the deficiencies being highlighted.'

To this day, Paul Lennon maintains he knew of Gunns' withdrawal only via the Australian Stock Exchange, but a year later, while under oath for an inquiry relating to government appointments, Simon Cooper refuted this. The lawyer told a Legislative Council committee in-camera that, 'He [Lennon] telephoned me on Monday, 12 March, and my note says, "Gunns to pull the pin this Wednesday."'

The day after Gunns pulled out of the RPDC, Lennon announced that an emergency cabinet meeting had decided to fast-track the assessment and his cabinet would appoint consultants to assess the project. The pulp mill's fate would then be decided by a parliamentary vote at the end of August – a deadline in effect set by Gunns.

The president of Tasmania's upper house, Don Wing, described the proposed legislation as 'Gunns' Dream Bill' and let it be known that lawyers for Gunns had helped prepare the bill, alleging that on the weekend before the new legislation was to go through parliament, the company's directors and lawyers met with the premier, deputy premier and senior government officials to put the final touches on the drafting of the bill. Neither Gunns nor state Labor has denied this. Ben Quin, a local federal Liberal Party candidate, was one of the few who spoke out against the new approval process of the pulp mill. His colleagues immediately attempted to silence him. When he later quit the Liberal Party to stand as an independent, Robin Gray called him a 'coward.'

Lennon appointed SWECO PIC, a Finnish pulp and paper consultant, to assess the health and environmental aspects of the mill, and ITS Global, an Australian consultancy firm, to oversee the economic and social benefits. Employees of SWECO PIC spent just two days on the island. Both consultants delivered their reports in just over five weeks. ITS Global declared the pulp mill would bring mostly high returns and positive social impacts – however, there could be some social tension in George Town for the duration of the mill's construction. SWECO PIC found Gunns had failed to meet eight of its 100 guidelines, in particular to provide proof that its effluent would not harm the marine life in the Bass Strait, but overall gave the pulp mill environmental and health clearance to proceed. 'The reports are not independent,' the Wilderness Society had warned. 'Whereas the RPDC had an obligation to consider the interests of Tasmanians, the two consultants merely had a commercial relationship with a client (the Tasmanian government) that wants the pulp mill to proceed.' The Forest Industries Association of Tasmania felt differently and declared, 'Tasmanians can now be assured that the mill is safe and can be excited about a higher standard of living to come.'

There are many who suspect the pulp mill had the state's consent even before the RPDC collapse. When Michael Hawkes, the former chauffeur for deputy premier Steve Kons, took the stand in a committee inquiry in 2009, he revealed that in early February 2007 he had dropped Kons at Paul Lennon's home for a meeting. 'The meeting went for probably about three-quarters of an hour, maybe an hour,' Hawkes recollected. 'And he came out and jumped in the car, and we drove off and Steve – I said, "How did you go?" and he said, "Okay." He was pretty quiet and we would have gone down the road maybe half a kilometre and he said, "The pulp mill will be approved by the end of May.'"

Not quite – but close enough. In August 2007, parliament was recalled a week early to vote on the pulp mill. It was approved. The act, as Richard Flanagan pointed out, explicitly ensures the mill will go ahead even if it is proven a consultant assessing the project has been bribed. When one member of parliament was later asked whether she had sought independent legal advice about the meaning of section 11 of the *Pulp Mill Assessment Act*, which states that 'a person is not entitled to appeal to a body or other person, court or tribunal … in respect of any action, decision, process, matter or thing arising out of or relating to any assessment or approval of the project under this Act,' she said she 'didn't have time.'

And that was that.

CHILE

After the RPDC process ended, John Gay, along with a cohort of state ministers, visited Chile to inspect the CELCO paper pulp plant, which was boasting its best ever financials, earning over US$600 million in the previous year. For Gay it was a positive experience. 'There is one pulp mill that has the wine growing beside its fences and the company is happily living with the wine industry. There is tourism there and all the people in the area are supporting the mill,' he said in a *Stateline* report. The state ministers were also in a positive frame of mind, giving CELCO a 'thumbs up,' with the sole exception of independent Ruth Forrest.

A few months before the delegation, Bob McMahon, a local teacher, rock climber and member of Tasmanians Against the Pulp Mill, had decided to take an 'unchaperoned' and self-funded trip to Chile to have a look at the elemental chlorine-free mill for himself. In particular he went to see a paper pulp plant in Valdivia that Gay and the ministers were soon to visit.

It was as if he and John Gay visited entirely different countries. McMahon returned with photographs of red and black graffiti reading 'MATA CELCO' (CELCO Kills) scrawled through the town of Valdivia. Locals showed him the purportedly thriving businesses brought to the area by the pulp mill: a tin shed that sold the odd beer, Coca-Cola and smokes to truck drivers, and accommodation built three kilometres from the mill, now abandoned and weed-infested. McMahon also learnt that the Chilean courts had shut down this Valdivian plant three times since its inception for environmental transgressions. When the owners announced their plan to discharge effluent into the sea instead of the river, there were violent protests from fishermen whom CELCO referred to as the 'natives.'

McMahon also met Dr Eduardo Jaramillo, professor of ecology and marine biology, the man in charge of the eco-toxicological monitoring program in the Rio Cruces Sanctuary, twenty-five kilometres downstream from the pulp mill. The sanctuary had once been the main breeding site in South America for the emblematic black-necked swan. But, as two studies affirmed, one by a biologist at the University of Santo Tomás and another by the National Forestry Commission, the reserve's swan population had plummeted from the thousands to a few hundred within months of the plant beginning its operations. The university found 8,000 swans had diminished to 518, while the commission found 6,000 swans were now 300. Locals had told McMahon they weren't overly concerned until swans started falling out of the sky and crashing through car windscreens.

Chile's national environmental agency concluded that the activities of the Valdivia plant set off a chain reaction that ultimately led to the swans' death. It said the birds died primarily from starvation because their main food source, a waterweed called luchecillo, is now almost completely absent from the bottom of the river. Scientists also found an unusually high concentration of iron in the livers and kidneys of the dead swans, which they believe is the result of any remaining luchecillo having an unusually high concentration of iron, copper and manganese. Thousands of residents took to the streets in protest against the factory. In 2004 CELCO was fined US$550,000 after the state determined run-off from the Valdivia plant had exceeded both temperature and heavy metal limits. The company refused to pay and appealed for the next three years. Temporarily shut down, the pulp mill was allowed to re-open in late 2005 on the condition it build a channel so that the liquid waste it had dumped into the Río Cruces River and nearby wetlands would be diverted to the Pacific Ocean instead, where the fishermen are.

Late in 2009 Chile had its first successful criminal prosecution for environmental crime under its new criminal code. The culprits? Two top executives working for CELCO at another one of its paper pulp plants. After thousands of fish and nearby livestock were found dead in and around the Mataquito River, local police decided to have a look around the Licancel CELCO plant. They discovered two clandestine and pollution-filled channels leading from the factory directly to the river. The discovery began a drawn-out process ending with the two executives being fined and ordered to provide two mini-buses and a soccer field for the nearby community. The legal arm of Chile's government then pursued a civil suit against CELCO, resulting in an out-of-court settlement, the wood pulp company agreeing to pay US$1.1 million to restore the river but still not admitting any legal culpability.

Oddly, the Tasmanian tour group did not meet with the professor. When Dr Jaramillo enquired why, the ministers said they had tried, but he wasn't in his office when they turned up at the university. 'We do have phones in Chile, you know,' was the professor's exasperated response. When Bob McMahon paid for Jaramillo's expenses to come to Tasmania and visit the island's key decision-makers, the forestry industry and ministry neatly quashed him. The Forest Industries Association of Tasmania put out a press release saying they were sorry about the supposed effects the Chilean mills had on his country and were happy Gunns would not be going about their pulp mill in the same way, while one minister attacked Dr Jaramillo's credentials and findings so blatantly and mistakenly that the professor and his university considered launching a defamation suit.

As for Chile's black-necked swans, the majority of the remaining Valdivian population have migrated to a lake further south, but are having difficulty reproducing because their new lake is too small and predators can easily access the colony's eggs. In 2008 only one couple reproduced successfully. A floating island is now being constructed for them in the hope they will incubate their eggs safely in the middle of the lake.

THE TOTTERING GIANT

I n my research I discover the writings of Evelyn Temple Emmett, the director of the Tasmanian Government Tourist Bureau from 1914 to 1941. In his 1952 book *Tasmania by Road and Track*, he writes:

> Perhaps some mathematician will work out the amount by which the taxation of Tasmania would be lessened if the island had not been sprinkled all over with Lost Endeavours. Roads beginning and ending nowhere and now growing crops of gum trees; tracks opened and obliterated by disuse; canals cut and smothered by sand; harbours built for ships that never used them; railways whose main work was to lose money; piers pushed into the sea to rot; bridges that invited floods to bear them away; shafts sunk where there was no metal; towns built and abandoned! Doubtless other countries have guided their inhabitants into blind alleys and left them there, but surely none can have a record in this respect to beat Tasmania's.

It is an eerily familiar sentiment.

You would expect that after five years, considering the 'unacceptable commercial delays' and the special legislation, the pulp mill would have been built. But Gunns needs a billion dollars to get the project off the ground, and ever since the ANZ Bank decided not to be a key financier, it has been without a partner. There have been potential candidates, suitors even, but no company or bank has yet been willing to commit.

Today, almost ten years after Gunns first announced plans for the pulp mill, all that stands at the proposed site is a boundary fence and patches of grass. The most activity the timber giant has managed on-site is a little mowing and vegetation removal (eagle-eyed protestors are swift to point out that the company didn't have a valid permit to undertake even this work). The knees of this 135-year-old timber company have begun to creak and along with it, the state's forest industry. But even so, there are numerous invisible strings helping to hold it up.

When Gunns momentarily 'shelved' the mill, citing insufficient funds due to the global financial crisis in late 2008, Paul Lennon, now the ex-premier, snapped at the Tasmanian public, 'I just hope that those people who have opposed the pulp mill can now find the alternative employment opportunities we are desperately going to need in this state.'

But despite all the threats of economic ruin, the sky had not yet fallen in on Tasmania. Gunns, however, seemed the worse for wear. In early 2010, the woodchip mills were closed on and off for weeks on end – as were a few sawmills. Contractors were forced to store their felled timber in the bush. Loggers called on the state to provide them with 'exit packages' and short-term interest-free loans, while lenders such as J.P. Morgan stated that Gunns was at risk of breaching its borrowing agreements.

In February 2010 the company revealed a 98 per cent drop in its half-year profit from $33 million to $420,000. Less than two months before the announcement John Gay had sold 3.4 million shares for approximately $3 million, raising suspicions that the chairman was aware of the company's impending plummet in profitability. In response to ASX enquiries, the Gunns board stated unanimously that they only became aware of their income collapse on the day before their figures were released – which, if true, is an appalling negligence. Morningstar's stockbroker Peter Warnes spoke for numerous shareholders when he wrote in his 24 February report, 'Confidence in management is shattered. What happened to continuous disclosure?'

To understand how a former stock-market darling could have such a dramatic fall from grace, we need to look at the short-lived wealth Gunns was dependent on. For much of its near decade-long success, starting in 2001 when it shot into the top 100 public companies index, Gunns had refused to dissect its income any further than is minimally required by the ASX, citing 'commercial in confidence.' Between 2001 and 2009, Gunns divided its business into only three segments, one of which, 'forest products,' included both woodchips and sawn products. But on the back of its recent profit slump, the timber company listed four segments in its half-yearly report ending in December 2009. For the first time, woodchips were distinguished from sawn timber.

John Lawrence, an accountant and former economist, took the opportunity to take a closer look at Gunns' statutory reports since 2001 and published the results in the *Tasmanian Times*. He estimated that woodchips, mostly from native forests, contributed 50 per cent of Gunns' revenue and 65 per cent of its profits over the previous eight years. He also found that 24 per cent of Gunns' total profits over the same period came from managed investment schemes (MIS), which at times were the company's lifeblood, one

year overtaking woodchips and forest products in their contribution to net profits. 'Wood-chipping and MIS have contributed
approximately 90 per cent of Gunns' profits over the past eight
years,' Lawrence stated. 'But without them the picture will be completely different.'

The federal government's managed investment schemes encour-
aged investors to make upfront, completely tax-deductible payments to MIS companies – covering seedlings, management fees
and rent for ten years – of amounts of up to $10,000 per hectare. As
establishment costs were only approximately $2,000 per hectare,
this gave MIS companies a windfall. Many of the prospectuses
from MIS companies promised lucrative returns, based on estimated growth rates and timber prices – but because of the rapid
growth of MIS plantations, site selection was compromised. In true
'Ponzi-scheme' style, returns to investors were propped up. In Tasmania especially, loopholes were neatly exploited, with the higher
than expected interest from investors leading to aggressive activities such as clearing healthy forests to make way for plantations,
despite the Forest Stewardship Council's policy that such ill-gotten
plantations could not be certified.

Timbercorp and Great Southern, two of the biggest MIS companies, collapsed in 2009, leading to major repercussions throughout
the MIS industry. Gunns has been keen to point out that it is running a very different, more sustainable show. 'We're an end user
looking for a resource, not a resource searching for an end user,'
Ian Blanden, Gunns' head of plantations, told the *Business Spectator*
in 2009. Which is true, if the pulp mill is built. Gunns, Blanden
also claimed, has a 'very diversified source of revenue, MIS making
up somewhere between 10 and 15 per cent of our annual revenue.'
Yes, John Lawrence agreed, Gunns was not solely reliant on the

managed investment schemes – but the suggestion that they were a minor part of the timber giant's business was, in his view, misleading. A percentage double Blanden's figure, Lawrence claimed, would be more accurate. In 2010 Gunns suspended MIS sales due to a widespread investor backlash against the schemes. With revenue from MIS suspended and native woodchipping under threat, Gunns had become a shadow of its former self – hence its continued need for a pulp mill.

And so the exceedingly rapid rise of Gunns could be attributed to landing on the tail-end of one industry sector, native forest woodchips, and then walking the red carpet to its successor – plantations.

And now?

'It's a perfect storm,' says Paul Oosting, head of the Wilderness Society's 'Stop the Pulp Mill' campaign. The combination of 'people power,' ranging from big business leaders such as Geoffrey Cousins to mass rallies attracting 10,000-plus crowds, headline-grabbing actions by activists, along with the strong Australian dollar bucking the trend in the global financial crisis, a gross decline in woodchip exports, wary financiers and shattered confidence in MIS plantation investments has kept the woodchipper's dream at bay. Added to these was Gunns' initial refusal to take the steps towards getting Forest Stewardship Certification standards, a globally recognised stamp of ethically sourced timber, which requires a social, environmental and industrial licence. Instead Australia's native-forest timber industry preferred to pen its own standard – the Australian Forestry Standard, for which Gunns was the first company to be certified, with Forestry Tasmania's own general manager on the AFS board. But globally, this standard hasn't washed with customers. That cut Gunns off from much of the high-end paper and pulp market; at the same time, it is unable to match the prices of cheaper uncertified competitors.

In 2009 a Swedish pulp and paper manufacturer, Sodra, told Gunns it would only consider financing an FSC-certified and plantation-fed pulp mill. It also announced it would prefer a totally chlorine-free process and wanted the alternative site at Hampshire to be reassessed. Then in March 2010, Japanese pulp and paper companies announced they would no longer be buying any Gunns products without FSC certification. Reluctantly, Forestry Tasmania and Gunns said they would seek certification. The biggest obstacle they face is that under FSC principles, any plantation converted from native forest after 1994 is ineligible. Having only agreed to stop this practice three years ago, Forestry Tasmania and Gunns may find that much of their plantation estate cannot be certified.

Gunns managed to get a small proportion of its woodchips approved as 'controlled wood' by the Forest Stewardship Council, having declared it would sort through its woodchips, removing material from high-conservation-value forests and other unacceptable sources, to do so. This is not full FSC accreditation, but approved 'controlled wood' can be included in products bearing the FSC's 'mixed sources' label. Gunns could get up to $189 per tonne for these graded woodchips, a slightly better price than for woodchips from older forests, which are poorer in quality and are a bit like the pulp version of seafood extender – cheap and not quite right. While global prices for woodchips and pulp were bouncing back, demand for native-forest woodchips was not, and this turning tide had placed commercial pressure on Gunns and FT to adapt. Increasingly, pulp and paper producers see native-forest woodchips as an inferior product, preferring single-species pulp from trees of a uniform age. Even China is now declining Tasmania's native-forest woodchips.

*

In March 2010 the Greens won the balance of power in Tasmania once more. They won five seats, the same number as in 1989, but now in a smaller parliament – the push to shrink the number of seats had come back to bite both major parties. And while the Greens have recently been tight-lipped regarding the pulp mill's future, their presence surely spells the end of political favours and blind-eyes.

Since 2006, when Premier Lennon won a third successive majority Labor government, Tasmanians have witnessed the collapse of the state's planning body, the resignation of countless scandal-ridden ministers, the closure of the Department of Environment, Arts, Parks and Heritage and the bizarre pulp mill legislation. They have seen a pulp mill 'info' bus driven around their streets, while agriculture, hospitals, the education system and the railways have gone to ruin. It's no wonder many voters ignored the joint press release put out by Robin Gray, Paul Lennon, Michael Field and Tony Rundle – two Liberal and two Labor ex-premiers – warning against a minority government with the Greens holding the balance of power.

This third Green–Labor accord went a step further than its predecessors. For the first time since the Greens formed forty years ago, two of its members were sworn directly into cabinet. To add to the jilted Liberals' horror, the lower house is literally getting a make-over, as Labor has agreed to pull apart the benches and desks currently lined up in the traditional face-off between the two major parties to segment the parliament into three.

For the Greens, it was a chance to prove that it's possible to play politics without sacrificing one's ideals. For the rest of parliament, it was a chance to test their much-loved theory that it's easy to take the high moral ground when one doesn't have any responsibility.

At Gunns headquarters, there was also a reshuffle. After its 'surprise' profit slump, major institutional investors such as Perpetual Trustees, IOOF Holdings, Concord Capital, Perennial Investment Partners and Schroders Investment Management called for an immediate change of management at the top, demanding that both John Gay and Robin Gray resign from the company. In March, Gay admitted to the *Examiner* that interstate investors were pressuring him to quit the board.

'These people said it would be better for the company if there were no Tasmanians on the board,' he claimed, painting himself as a local patriot. 'But if I go, the mill, the Launceston head office and our interests would go with me.'

A month later, Gunns announced a restructure. Investors and activists held their breaths, waiting for the scalps of Gay and Gray. The company introduced the 'Southern Star Corporation,' a new subsidiary to look after the pulp mill and plantations, while Gunns would retain its sawmill sector and hold a 51 per cent stake in the new business. John Gay would resign as head of Gunns – but who was to be at the helm of Southern Star? John Gay, of course. Robin Gray resigned from the board, but would remain director of plantations, Gunns' biggest asset.

Shareholders and campaigners were furious. 'It's just the same old sleight of hand, very bad corporate governance that Gunns have become famous for,' said Geoffrey Cousins. Gunns' largest shareholders, Perpetual and Concord Capital, said the superficial reshuffle was unacceptable. Perpetual and IOOF offloaded about

1 per cent of their stake. Both the Commonwealth Bank and the Bank of America ceased to be major shareholders, and Gunns' share price dropped to 26.5 cents (Gay had sold his shares at 90 cents a piece in December). IMF, a litigation company, announced it was considering a class action on behalf of disgruntled shareholders, who allege that the company breached its market-disclosure obligations. The company sold its hardware stores and put 28,000 hectares of private native forest on the market – but it was only in May 2010, with the complete and abrupt departure of Gay and Gray from both Gunns and Southern Star, that the share price steadied. Nearly 40 per cent of the company changed hands in the last week of May. Speculation about a take-over nudged the value of the shares up, but gone were the glory days of eight-dollar, sometimes twelve-dollar, shares. Silently, the names of John Gay and Robin Gray slipped off the Gunns and Australian Stock Exchange websites.

BASEBALL-BAT DIPLOMACY

'I t's the way they go about things,' Bruce Montgomery says to me, recalling a run-in he had with the company. 'From the very beginning they went about things the wrong way.' Montgomery has a long and not altogether happy history with Gunns. While working as the *Australian*'s Tasmanian correspondent in 2003, he accepted an offer from Paul Lennon to join the Forest Industry Council, including the newly established Pulp Mill Taskforce, as communications manager.

'I'd said I'll only do it if I can achieve change,' laughs Montgomery, remembering his conversation with Lennon. 'I was going through this paranoia that all isolated correspondents seem to go through, that they're being forgotten. Plus I was driven by this desire to try and end the war. To resolve the forestry conflict.'

As a young reporter in the 1980s, Montgomery had led the charge against the Wesley Vale pulp mill. 'I used to have enormous fights with Gray over that pulp mill,' he says. Now, he found himself part of what was essentially a government drive to convince the public that a pulp mill was a good idea. 'For good or bad, it was to promote the idea of a pulp mill without promoting Gunns' pulp mill,' he says. 'It was supposed to be an educational program, then it became propaganda.' But Montgomery was determined not to let bias cloud his judgment. He learnt everything he could about modern pulp mills, wanting to find out if the technology had improved since the days of Wesley Vale. 'The conclusion I came to was that it has – you can have safe pulp mills around the world now, providing you do it properly.' He pauses. 'My fear was not that you couldn't have a safe pulp mill, but if Gunns would do the job properly.' He spent eighteen months in the job before resigning, exhausted and frustrated by the industry's refusal to take his advice seriously.

Montgomery's first encounter with Gunns, however, was not a professional meeting but a neighbourly dispute in the late 1970s. He had bought a house in Launceston. 'It was a village-green residential street lined with terrace houses on either side and a Gunns timber yard down one end.' At the time, Gunns was still run as a family business headed up by David Gunn, a former mayor of Launceston. John Gay was general manager. 'I was doing up my terrace house Balmain-style and Gunns decided they were going to develop their yard into a big hardware outlet by buying up the street and knocking it down ... They were creeping around at night and scaring little old ladies to sell their homes, saying things like "You know it's going to happen, nothing you can do, so take $15,000 and move on."'

Montgomery set up a local action group to protect the heritage street. 'We eventually got an agreement with the Launceston city council that if Gunns developed their hardware store in Simater Street, as soon as a new road system was developed at the back of the property, they'd flip the development to face that way and the Glebe precinct would be returned to its English glory.' He shakes his head. 'That's never happened. A new road is there, but they never followed through. Every time I return to Launceston, I go to look at our old home and see them and think, you pack of bastards, we were conned.'

*

In May 2010, Premier Bartlett initiated a forestry roundtable to discuss the future of the island's timber operations. A $3.6 million logging contractors' assistance package was announced, which added to the $1.8 million already committed, during the recent state election campaign, to help loggers pay wages, interest charges and equipment repayments while the woodchip mills remained closed. Loggers continued to call for a government-funded 'exit package' to allow up to 60 per cent of their number to leave the industry with dignity, not as bankrupts. In July 2010, the federal Labor government offered to provide financial assistance should the roundtable result in an agreement.

When he announced the roundtable, Premier Bartlett commented that blaming environmental groups for the downturn in woodchipping was a 'reasonable conclusion' to draw. Sue Neales, a senior journalist at the *Mercury*, responded by suggesting the premier ask some crucial questions of the state's timber industry instead of laying the blame on green groups:

> Why did the local industry ignore the clear signals apparent more than five years ago that international demand for native-forest woodchips was on the wane? ... Why did the industry not move more quickly into reliance on plantation woodchips during the mid-2000s when it was obvious that a woodchip export pile comprising mixed tree species of variable ages logged from Tasmanian native forests would never be as cheap or productive to process into pulp as single-species, uniform-age eucalypt plantations maturing across vast tracts of South America and South Africa? ... What has happened to all the tens of millions of federal dollars awarded to Tasmanian timber companies since 2005 to help them retool, adjust and move to a new, more sustainable

future under the Community Forest Agreement, when the sector is now so clearly in trouble and in need of real change?

In a *Four Corners* investigation into local reactions to the pulp mill in 2007, reporter Liz Jackson asked a scalloper, John Hammond, why he thought the government would want to damage the fishing industry. Hammond said he didn't think it was malicious, just dumb. 'They're dumb. They don't understand. They don't want to understand. They've made a position, they've taken a position and now they get the baseball bats out and fight anyone that argues with them. We've got baseball-bat diplomacy taking place and that's just the way this government operates.'

But if this is true, then what makes Tasmania so vulnerable to such blinkered visionaries? And how does this affect the psyche of ordinary Tasmanians? After reading a draft of this book, in particular the section on the thwarted RPDC, a Tasmanian writer tells me she fell into a depressed sleep, waking up intermittently from angry dreams. 'The rage came on me unexpectedly,' she writes in an email, and cautions me that reading all the details amassed in the one document could have 'the effect of shattering those coping mechanisms that some of us here employ in order to stay sane.'

I am unsure how to respond. This is not something I expected when I came to Tasmania to see a forest before it was logged. I never considered that the feeling of theft and betrayal would extend so much further than the fluoro-ribboned tags outlining the logging coupe.

Each time I've sat down to type out the details about the pulp mill assessment, Gunns, Lennon and the complacency of state parliament, I've felt sick. It's so blatant. When I asked Paul Lennon why he chose to fast-track the pulp mill, he responded simply,

'The fact is that if the RPDC were left with the project, the mill would be dead and built in China.' That's not a very helpful answer, but when I look at Lennon, I know he thinks he did the right thing. That somehow – even though the figures won't balance, an independent process perhaps already fragile from years of political leaning has crumbled, men have worked themselves into a state of righteous violence, towns have split, people's lives and businesses have been put on hold – somehow, in the flaming furious inarticulate world of Lennon, he's done the right thing. I cannot match the former premier's certainty. I agree a pulp mill might have been a success, a nice little earner, but not like this. Perhaps if the democratic process wasn't considered so pestilent in the state of Tasmania, I could see the pulp mill through Lennon's eyes.

Documents keep surfacing. In 2009 on the *7.30 Report*, reporter Conor Duffy revealed secret papers showing that Julian Green of the RPDC had written to John Gay as early as 2005 to convey his concerns about odour potentially escaping from the mill. Dr Raverty told Duffy he and Green had visited Sweden during the assessment process to observe the 'world's best-practice pulp mills.' 'When we got out of the minibus in the car park, Julian Green very quickly became distressed – he couldn't breathe. I found the odour intensely objectionable, and within a matter of minutes, Julian Green was gasping and saying "For God's sake, get me out of here."' Raverty said the panel was very concerned about fugitive emissions.

In the same year a report on the pulp mill by CSIRO's leading oceanographer, Dr Mike Herzfeld, was finally made public by Gunns after months of resisting its release. Among several damaging conclusions, the report found that the mill would breach water pollution levels on an almost daily basis. Gunns, however, quickly followed up the report's release, as they did the *7.30 Report*'s findings, by claiming that new technology has dealt with these concerns.

As it stands, if the pulp mill as proposed and legislated is built in Tasmania, its 64,000 tonnes of effluent released daily into the Bass Strait may, or may not, affect the yearly passing of endangered humpback whales. Salmon farmers and fishermen may find nothing unusual happens to their fish or quota. Or it may. The valley could smell like rotten eggs and boiled cabbage and the grapes on the vine bloat like foul farts. Or not. Local wine may keep winning medals to add to the round stickers on their bottles or the judges may begin to detect a strange, metallic taste. Of the forests in the north and north-east, and the moss forests and dry bush outside reserves, all may be required to satisfy the hunger of the mill, the chippers and the wood-fire burners. The seal colony 15 kilometres from the outfall pipe will be fine. Maybe. Of the thousand-odd timber workers who have lost their jobs in the past decade, ten or a hundred may be employed at the mill. The local sandwich shop in George Town may do good business, but then again, the mill could have its own canteen. Of the 26 to 40 giga-litres of water allocated to the mill, operators are hoping they'll use less. But it's difficult to be certain. If the independent planning body hadn't been bullied and dismissed, many of these questions could have been answered. And Tasmania might have had a pulp mill by now. Or not.

*

The future of Gunns is uncertain. The FSC accreditation process is a lengthy one, typically taking two years. With \$600 million of debt, Gunns doesn't necessarily have that kind of time up its sleeve. It is said buyers are circling the former timber giant's plantations.

A pulp mill is still on the cards, according to the company's management. Conservationists, on the other hand, say persuasively that the mill as proposed by Gay is dead and that it is the 'beginning of the end' of logging in native forests.

There are signs that the culture at Gunns may be changing.

'How do you feel about protected species dying for your business?' John Gay was once asked on national television.

'Well, there's too many of them, and we need to keep them at a reasonable level,' came his reply.

In June this year, however, Gunns announced it would no longer use 1080 poison. The new CEO, Greg L'Estrange, described this decision as 'another significant move in Gunns demonstrating its social responsibility.'

Meanwhile, logging continues. Over 193 coupes containing old-growth forest were earmarked for clearing in Forestry Tasmania's 2007 three-year wood production schedule. None of this has ceased. I get the sense you could turn your back on the ailing timber giant, only for it to gasp back into life, resurrect itself as 'Gay & Gray' and rattle hungrily back into the woods.

*

Back in 2008, John Gay scoffed at rumours ANZ might withdraw financing for the pulp mill. 'ANZ and Gunns deal on a commercial basis, not on an emotional one,' he had said. 'I don't talk to financiers about environmental issues and they don't raise them with me.' Several weeks later, ANZ dropped out of the deal. Gay seriously underestimated his opposition. What was even more telling was his assumption that environmental issues are emotional, not commercial. In 1997 the top scientific journal *Nature* published a report by a team of economists and scientists from the United

States, Argentina and the Netherlands in which they calculated the value of services provided to humanity by the natural environment at US$33 trillion per year. Nature in Tasmania has no official administrator, receiver or liquidator to appoint. To date, all it has been able to rely on is a motley group of feral renegades, suited dissenters, intuitive battlers and educated professionals to hold the line.

Beyond all the figures, statistics and sums, the question remains: why are – were – John Gay, Robin Gray and their ilk so obsessed with the pulp mill? Why the strange phone calls to Gunns 20 defendants, Liberal leaders and a celebrity gardener? Why the bizarre opinion pieces? For an issue that they say is purely economic, why does it seem so *emotional*?

I am reminded of the hotelier and creator of the Ritz Hotels, César Ritz, who, after failing to pull off an ambitious project, deteriorated mentally and spent his last years in an asylum, where he drew the same thing over and over. Hotels. Will John Gay do the same thing now that the pulp mill as he envisaged it is dead? When Robin Gray was asked to paint a picture to auction off for charity three years ago, he painted five faceless Greens candidates holding yellow placards at Salamanca Place. He called it *Tree Huggers' Paradise*. Is it possible that, beyond all the carefully laid out arguments and the picking over of Gunns' financial carcass, these men are fuelled by something as simple, base and merciless – and so like a Grimms' fairytale – as hate?

EPILOGUE

I make friends with a girl who has nothing to do with trees. She's a rare find. She takes me out to the coast where she and her sisters used to go to find shards of porcelain plates and teacups and saucers. It is a bit of a mystery why the hill and rock pools and dunes are filled with broken crockery. She thinks it had to do with a bacteria, a disease maybe, that prompted settlers to walk their favourite tea-sets and dining plates down to the water and smash them on the rocks before burying the fragments deep in the soil so people wouldn't smuggle them back into the colony and start the germ up all over again. When we get there, someone has built a huge sandstone gate where once one could walk down to the beach. We backtrack a little, look for electric wires and step under an ordinary paddock fence. Following it down to the sea, we studiously ignore the 'Trespassers Keep Out' sign posted in the dirt.

Clambering over the rocks and smelly piles of kelp we start to see the small bits of china. Softened and aged by the salt water, the broken pieces are smooth like skimming stones. Quietly we pick up pieces, turn them over and show each other the patterns: pale blue ink that reminds me of prison tattoos, cursive loops of flowers, Ming china waves and rouge autumn-leaf prints. We transfer some fragments with sharp edges to deeper rock pools to soften, while further up the dune I find a porcelain piece with a bird's leg and its feathery underbelly printed on it. Finally we sit on a rock to lay out our finds and help each other pick out our top three pieces. One of mine is a piece of dimpled purple glass, frosted with salt. My grandma used to collect bits of softened glass on the beach. She would have loved a purple shard. At the end of the day we finish fossicking and tread back along the kelp now heaving with the incoming tide. The broken plates looked like shells, just as curious and gentle, not like they don't belong at all.

It is a relief to find beautiful traces of us.

*

Camp Florentine has been busted. At a meeting in the Pink Palace's kitchen, Wazza looks over his handwritten agenda before looking up at the gathered crew.

'Should we start with the "They're smashing us to bits" section?' People nod. He is about to keep talking when a couple of new arrivals stumble in the back door. They've just come from the police station.

'I'm sorry,' one of them says. 'I didn't mean to get arrested. I'm really sorry.' The crew is running out of arrestables: 'bunnies' without bail conditions are precious cargo right now. 'They just drove into us, cut us off,' he continues. I think he means the police, but I learn later that it was Forestry Tasmania vehicles. 'Nat's still in there. They're charging her with assault.'

'Wha–?' some of the crew in the kitchen exclaim.

'Yeah, she flicked a leech onto [FT employee] Marriott's hand.' Everyone laughs. 'But seriously, Marriott wants her done with assault.'

'What about the footage? Did you get footage?'

'Yeah, Ali's got it. But he's at the cop station too. He was already on bail, remember?' People breathe out, others are thinking hard. A sea of dirty dark hoodies, all accustomed to talk of the cops and bail and bunnies.

Almost a year later, most of the activists' recent charges will be dropped after their defence counsel discovers Forestry Tasmania wrongly advised police of the exclusion zones. The evidence sitting in Ali's pocket in a police cell also comes in handy at a future court date, when the footage contradicts Forestry Tasmania employees' claims that they were assaulted by activists. Instead the onus is reversed, with a nuisance charge laid against the FT workers: the video evidence shows they used their cars to cut the protesters off. But this evening no one knows that. Instead the blockaders study their bail conditions for loopholes, for a place they can lock on and stop the timber getting out of the Florentine for at least a few more hours.

*

Liesel is cutting up vegetables and putting them in a bowl.

'It's like that saying,' she tells me. 'If you put a frog or a duck in boiling water, they'll jump out, but if you put them in cold water in a saucepan and slowly boil the water, they'll be cooked before they know to escape.' She looks up and smiles, cucumber mid-cut. 'I usually use it as an analogy for bad relationships, but it works well for climate change too.' As she puts the cucumber slices into a bowl, Wazza takes them out, crunching happily. So far Liesel hasn't noticed that she is the Sisyphus of chopping up vegetables. As I settle down for the night, sharing the lounge room with three others, I wonder, with the reality of global warming becoming more accepted, understood and urgent, if these activists are subconsciously demonstrating the need for a new kind of job?

Years earlier I met a chemist who worked in Antarctica on a project to clean up toxic waste left behind by earlier expeditions. She was a glamorous cleaner, we joked, and there were times at the blockade when I watched the activists work and thought they were like slighter, gentler shadows of the loggers, enjoying the physical exertion as they worked outdoors to build obstacles, restore bull-dozed tracks, record species, rehabilitate silt-filled streams and make temporary bridges. But none seemed to consider it their right to have a paid, 'proper' job. When the original Florentine activists received a grant to restore the Churchill Hut, they were pathetically thankful for the money, although it barely covered their costs, let alone wages. Why have their rights to work and to put forward new ideas been ignored, while the loss of 300-odd timber-felling jobs is treated like the sky falling in? Is this why the activists have put themselves on the fringes of society?

'It breaks my heart that we can't offer paid work,' says Todd Dudley, who is running a restoration project along the Skyline Tier on the east coast, returning 280 hectares of pine plantation to native forest. 'I had one local guy who had volunteered here and wanted to keep working, but we couldn't pay him a wage. So he had to go back into logging.' A conservationist with twenty-five years' experience in land regeneration, Dudley thinks jobs in restoration provide the best hope of reconciling the timber workers and the environmental movement. 'It's strong, outdoorsy and smart work. You can employ both sides.' The land at Skyline Tiers was in poor shape when he came across it; the soil would barely have been able to support another plantation. But it took him four years to con-vince the company leasing it, Timberlands, to regenerate rather than plant more pines.

The project could provide at least thirty part-time jobs, but the past three years have seen numerous volunteers from conservation groups, the local community, people fulfilling welfare obligations and a Greencorps team (ten local seventeen to twenty-year-olds) working on the land, pulling out weeds, using machinery to remove renegade pine trees, replanting eucalypts and native grasses, and rehabilitating the streams that feed the surrounding catchment, lagoons and estuaries. In economic terms, the land as Dudley found it was decreasing in value, but with care it is slowly coming back to life. 'It's a good story in among all the rest of it,' he says.

I think of Forestry Tasmania's finances, the decreasing value of its 'biological assets.' Surely there are jobs to be found in restoring it? Or will the idea that environmental protection and economic growth are incompatible persist?

*

The next morning I go out to the Florentine Valley. In Maydena, Prue had painted on her fence 'Show Us the Maps,' a reference to Forestry Tasmania's claim that 90 per cent of the Upper Florentine is 'protected or otherwise unavailable for logging.' Up until now, all requests to see the maps proving this statistic have been ignored. In her living room Prue is trying to crack the code. 'I'm even looking at my grandfather's maps and 90 per cent just won't make sense.' She is surrounded by old and beautifully hand-drawn maps. 'It doesn't matter if you're a logger or a greenie,' she says, 'it's the fact that our government thinks its electorate are a bunch of dimwits.' So far her fence has survived.

When I turn the corner and drive up the straight stretch to where the blockade was, there's a muddy temporary camp on the side of the road. A cluster of people are standing where the logging road shoots into the forest, a strip of plastic cordoning them off.

Police stand in front of it. Parking near the new camp, I walk towards the disparate group and spy Miranda, kneeling on the bitumen a little apart from the group, her hair like a curtain around her face. Biting her fingernails, she's looking intently up the logging road. I hear a chainsaw and catch a glimpse of a man in a fluoro vest. Then I realise what's going on. Her tree is going down.

A few local journalists arrive and the police inspector comes to the mouth of the road to escort them through. One reporter still has the plastic price tags on his Blundstone boots and a couple of the young female journalists are wearing high heels. They totter up the hill with the inspector. It is quiet for a moment. Then the buzz of the chainsaw starts up again until, finally, a crack. It is a big tree, some 70 metres tall and about 300 years old. When it starts to fall, I can hear its canopy swish, falling into the arms of the surrounding forest, catching it like an office 'trust game' before it falls through, taking other smaller trees down with it to the ground.

I look at Miranda. I don't know how to comfort someone who has just lost a tree. I also don't want to intrude. She looks up at me, her eyes dry and thoughtful. She gives me a weak smile and mouths, 'Hi.'

Thirty police walk up and down the road alongside trucks and excavators to repel any attempts by the activists to lock on. There's a police bus with a satellite dish on top and portable loos. Workers have dismantled the activists' blockade and tossed it into a nearby quarry. Afterwards, Forestry Tasmania took photos of the quarry and released them to the media with the caption 'ferals rubbishing our forests.'

At the makeshift camp on the side of the road there is a battle-weariness – there are some new faces, but all are tired. There is a sense of surrender to the mess and grit of their circumstances.

One young guy says they can't get dry. Everything is damp. Their feet look like muddied zeppelins, boots bloated with clay.

Down the road, Petal is holding up the gravel trucks. Petal is the fellow who hitchhiked down the highway with a plastic pirate's sword. He is perched in a tree, locked on to cables and wooden poles, blocking the way to the quarry. A couple of trucks and their drivers grizzle. There are a few black wallabies in the bush, I am told, secret ground crew to keep an eye on him. Everyone else has been cleared off. I manage to get permission to stay. Beneath Petal the search-and-rescue police are unwinding their own ropes and cables, securing him so they can shift the roadblock. When Petal stands up on his branch, his movement pricks the ears of the police below him.

'You coming down?' they yell eagerly up at him. He reaches into his rucksack, which is nestled in an armpit of the tree. A family of ringtails stare at him from the hollow.

'Nope. Just getting some food,' he hollers back. 'You want some?'

You can almost fall in love with the stubborn childishness of it.

*

I return when the log trucks, police, satellite dish, security guards, forestry officials, gravel haulers have gone. The yellow plastic exclusion tape has been torn down and flaps on either side of the logging road. I say a quick hello to the blockaders, still in their temporary camp on the side of the road, before disappearing up the road. I don't want anyone to come with me. I want to gauge how I feel as honestly as possible. The wooden tripods, khaki banners and Miranda's tree are all gone. The man-ferns the blockaders had propped up are flattened again. I walk up the gravel road talking to myself – 'Oh, it's not so bad, it's not so bad. That'll grow back, it's not that bad, look, that's still here.' Then I turn the corner as the gravel doglegs away from view of the main road.

It's gone. It's all gone. It feels as though I'm at the edge of the world. The loggers were only in here a little over a month. Logistically it is incredible: a vast forest, thousands of years old, gone in six weeks. The gravel widens into the size of a two-lane highway and tapers off in a few directions. So far, two kilometres have been pushed through. There are small bits of blue wire in the ground from the dynamite and odd inky puddles in the tyre trenches. The trees left behind are precarious. They creak in the wind. One blockader told me she could hear the lone leftover trees cracking and falling over during the night, unable to withstand the wind without the surrounding forest. Ferns have been lopped and some sit upside down in the mud.

There are three cleared sections adjacent to the road and I climb over the stacks of debris to see them. The upturned roots and mangled logs are piled up high; beyond them is a kind of muddy swamp. I don't know how deep the mud is. Carefully I crawl and shimmy over the stumps. When I get to the edge I cling to a log like a leech and wriggle along it until I can see. Again, it's all gone, a stadium-sized pile of mud, rubble and nothing.

On my way back, I slip.

I sink into the grey mud, the wet oozing inside my boots, socks and jeans. I have to clench my toes to keep my boots on as I hold onto a branch to pull myself out. The branch snaps and one boot comes off. I plunge my foot back down into it. The mud makes slurping sounds. I feel like I am in the Swamp of Sadness from *The Never-ending Story*, the wet sucking me down, its sadness swallowing me. I grab another branch and wrestle my legs out of the mud. The swamp pops as it releases me. I'm excruciatingly slow making my way back to the logging road, hauling myself up over the logs. Other than the clumsy slipping and sliding of my legs and heavy breathing, there is no sound. I feel like I am the only live scrambling thing in here. I step back onto the gravel road, dragging the stinky clay guts of the forest with me.

*

For four years this blockade has been pressing pause – six hours there, thirty-four hours here, another ten hours, week by week. In the depths of winter, wearing wetsuits up tree-sits, sleeping in shifts, endlessly scouting, listening to the beer-sloshed rebel drawl of Redgum and digging shit-pits: it is a rough and ready world. But doubts do creep in.

'I'm not necessarily against native-forest logging,' says Wazza. 'I said that to a contractor once – said that if they protected this one bit of forest over there, I'd happily go home and you can keep logging. And he replied that's all very nice, but then another me would come along and ask the same thing. He got me there.'

Miranda is also thinking about what their crew wants to achieve. Is it to rescue one forest and let the others go to the chipper? 'We could end up protecting the Upper Florentine. It would be easy for them [the government] but risky for us.' When I ask why, she is quiet for a moment. 'It can't be about playing off land for land anymore. A reserve agenda is incremental. The industry needs to be reformed. The validity of wood-chipping needs to be questioned.'

Miranda and I are sitting on a jetty in Hobart. A couple of albatrosses are eyeing us warily from the far end. Since her tree came down, she has been living in a share house with six others, her first rental home in four years.

'It was strange at first,' she says, 'I had gotten used to living out there. But it is kinda nice now.' Miranda and Nish are due in court soon over the assault charges against the loggers who allegedly attacked them.

'A man from the contractors' association has been in contact with us about it. He wants Rod Howells, the guy who did it, and us to come to an agreement outside of court. He asked to meet with us at Salamanca to talk about it.' Wary, Miranda talked it over with Ula and Nish. They agreed to meet the spokesman for a coffee. When they arrived, Rod Howells was sitting at the table too. Miranda says she shrank back.

'I didn't realise he was actually bringing Rod to the meeting,' she tells me. 'I just went quiet. I couldn't really speak. It's like they still didn't get how scary that was for us, like it was nothing.

'The man who had contacted us talked about Rod and that he had lost his daughter and the anniversary of her death was coming up.' Miranda looks down, fiddling with the teabag in her cup of tea, most of which has spilt over the jetty into the water below. 'I don't know,' she says. 'I'm someone's daughter too. That's what my mum kept saying to me when I told her about the meeting.' It was difficult for the blockaders to gauge if Howells was actually sorry and a bit inarticulate, or worried he wasn't going to get off the assault charges. 'He kept saying to us that he hasn't been bothered by protesters since the incident, so you know, it worked.'

Looking up at me, she stops trying to direct the spilt tea away from us and reads the look on my face. I'm furious for her.

'Still,' she quickly adds, 'the fact that we all sat together at a Salamanca café, that's amazing. Some of the workers at the action that day had threatened us, swiping at their throats and said that if they ever saw us in town, they'd cut our throats. And here we were, sipping coffee.'

I nod, trying to match her hopefulness. But the fact that no one even considered that it might be intense for Miranda to come face to face with Howells makes me angry.

The day after they all drank coffee together, they stood in the foyer of the magistrates court. Rod Howells stood alone near the window. He had combed his blond helmet of hair flat over the collar of his royal-blue suit. A small group of activists sat in a huddle on the other side of the room. The defence and prosecution nodded at each other, and Miranda whispered a 'hello,' but in a sense it was back to the trenches.

Inside, Howells' lawyers argued that the police evidence – video footage of the incident – wasn't valid because it had travelled from hand to hand and been in various machines before it reached the police. The girl in the tree who filmed the attack had given the tape to another activist, who drove it into town and gave it to Ula, who copied it onto her computer, burnt it onto a CD and gave it to the police. The police then edited the footage, some twenty minutes' worth, down to the crucial two minutes.

Outside the court, activists whispered to one another about who had the original tape.

'It was in a shoebox,' said Ula. 'And I gave it to ———— for safe-keeping.' After a few hours, the case was postponed for three months.

'It was weird,' Miranda says. 'Rod came out of court and sort of smirked at us.'

*

With 500 others, I return to the Upper Florentine. A public rally has been organised by Still Wild Still Threatened and other environmental groups.

Some locals have painted 'Save a Job, Shoot a Green' on the rock face next to the road leading into the valley. Three carloads of youths stand to the side of the rally, drinking beer and mimicking protesters, one wearing a Rastafarian wig of long black dreadlocks. Later, when one throws a rock at a girl's face, she complains to the cop on the side of the road. He gives her a steady look.

'Well, did you just trespass into the coupe?'

'Maybe,' she replies.

'So it's okay for you to break the law and not someone else?'

Unlike previous rallies, where there have been two lines of cops, this time there are only a few police officers at the mouth of the logging road. The rally nudges up against them. One officer announces through a loudspeaker that it is illegal to enter the site and people will be charged with trespass if they cross the line. Another holds a video camera, panning back and forth over the crowd. I recognise some faces – lawyers, scientists, photographers, the fezza mob, some Green politicians – but everyone is becoming one pushy mass. I feel the lens of the camera flick past me and then stop, pan back and hover on my face for a few seconds. My skin prickles.

During my time on the island, I've been getting increasingly paranoid. The case against the last four Gunns 20 defendants was dropped in February, with Gunns agreeing to pay costs, but one of them has mentally unravelled. A filmmaker, he had been beaten on the side of the road by a logger, Gary Coad, falling backwards into the gutter and damaging his back. Today he will no longer answer his phone or open his front door – a twitch in the curtains is the only sign he is home. At night he is spooked by spotlights, sometimes real and sometimes hallucinated, shone in his windows by bullying locals. Another of the Gunns 20, Heidi Douglas, told me the case made her absurdly paranoid for a while. The writ against the defendants referred to photographs of them coming out of meetings together. It appeared they'd been watched, perhaps by private investigators.

'When I heard that,' Heidi said, 'I immediately remembered the time this guy was snapping photos of the steps a couple of us were standing on near the waterfront. We were having a chat and it had struck us as odd, but he was dressed like a tourist so we let it go.' Heidi says she started to keep her curtains drawn and became hyper-aware of anyone taking photos. 'There was a brief moment when I thought there might be people hiding in the bushes outside where I lived!'

Some time after Heidi told me that story, I met with Lindsay Tuffin, the editor of the news website *Tasmanian Times*. We had ducked into a newsagent together in search of a book and as we left the shop, a man dressed in tourist garb – polar-fleece vest, hiking boots, Akubra hat and reflective sunglasses – stood in front of us, taking photos of the doorway. I turned around to see what he was photographing and saw only the counter with its shiny packets of gum and magazines. And us. I went cold. I wanted to ask who he was, what he was photographing, but Tuffin was already heading off towards a bookshop, saying we'd have better luck there.

I pushed the suspicion away, putting it down to an inflated sense of self-importance. But the paranoia was there to stay. Tinges of it followed me all over the island.

And the idea is not as outrageous as all that. The smallness of the island results in bizarre crossovers and cameos. Gemma Tillack, from the Wilderness Society, tells me about a conga-line of enemies driving up the Midland Highway to Launceston, all going to meet with federal ministers.

'Our car, which was full of forest activists, overtook the car in front and as we passed we saw it was being driven by Barry Chipman and was loaded up with timber workers. There was a log truck hauling old-growth timber in front of us. The three of us were stuck like that for the entire drive up the middle of the island,' she laughs.

I got a shock once when, while I was travelling in a logger's car, a wheezing old Datsun pulled up along side us at a red light, carrying a bunch of Pink Palace crew. Carefully I pressed myself back into my seat and turned my face away, feeling like an adulterer but not sure who I was cheating on.

And so, caught in the spotlight of the police camera, I feel trapped. This whole time, I've been hyper-aware of my instinctive affection for nature; I can't walk past a cat or a dog or a mudlark without stopping to coo at it. I live in a small flat with a tiny balcony, yet plants drape over the edge, almost touching the ground one floor below. I've tried to balance my seesaw heart, carefully weighing up each argument. But there is something about this island that wants you to choose sides.

I pull my gaze away from the video camera. The policeman repeats that they will arrest anyone who enters the forest.

'But this is state forest!' someone screams. The crowd is like cattle dogs and sheep in the one body, we are pushing and hesitant all at once, as though unsure who is leading and who is following. I see Pete Hay, his piercing blue eyes, white hair and weathered fisherman's face swaying with the push and pull. I catch his eye, we grin, and inside I say, fuck it.

My hands flat against the back of the person in front of me, someone else's hands on mine, I feel the crowd start to move. We surge over the invisible line and into the forest. The forest has been cleared and pushed to the sides. When we reach the point where the road turns and widens, there is a pause, a gasp, then people keep going, up and then down the hill towards two excavators.

The drivers have turned off their engines. Both stay inside their cabins. One is playing 'I'm Proud To Be a Redneck' at full volume.

> I'm proud to be a redneck
> you know, life is pretty sweet ...
> Got an outhouse and a yard
> ... real good family history too

But the song is soon drowned out by the approaching chant: *Our Forests, Not Gunns', Our Forests, Not Gunns'.*

A helicopter flies over, a photographer leaning out. Whenever a policeman tries to talk through a megaphone, the chanting gets louder until he gives up. I wander over to the machine furthest down the road. The driver, sitting inside with the door open, is pale and pudgy like a sponge cake, with blond hair and round blue eyes. I ask him if he's okay. He nods, says he doesn't care. 'I'm getting paid anyway.' Twenty-six years old, Matthew has a mortgage to pay, as well as child-support payments to the mother of his two children. He left school when he was fifteen. No one in his family seemed to mind. 'My dad didn't finish school and nor did his dad. Didn't like it.' We start talking about his kids, and how his grandfather was a logger.

A Forestry employee comes over and gives me a look.

'Everything all right here?' he asks Matthew. 'You okay?' Matthew nods. The man gazes at me before deciding to leave it at that and walks away.

Later, when the rally is over, I'm standing next to my car when a truck comes out of the coupe and starts down the main road. I see Matthew in the passenger seat. Instinctively, we wave.

*

'Why Tasmania?' Barry Chipman once asked me. He's right – in the greater scheme of things, the island is nothing but a drop in the ocean. But its story is universal – and what goes on in Tasmania goes on in the mainland, goes on in the Pacific islands, in other continents, until it comes straight back over the ice to Tasmania again. You can follow its story like a ball of wool, get tangled in it and unravel it.

One year ago, a fridge-sized container was found bobbing in the Tasman Sea. It was full of ammonia coolant and was traced back to a couple of astronauts who had turfed it out of their spacecraft. Where did they think it was going to go? Did they imagine it would levitate for the rest of its existence? Or that with a 70 per cent chance it would land in the ocean, it would be okay? Deep down in our bones we *must* know – we must know that nothing we do is done in isolation. Cause and effect: how did it get so noisy in between?

Some scientists are beginning to describe the modern geological era as the Anthropocene, the sixth in a series of mass extinctions, all said to be caused by extreme phenomena, in this case the harmful activities of humans. Perhaps even more poignant is biologist Edward O. Wilson's description of the period that will follow. Wilson says it will be 'the Age of Loneliness' – a planet inhabited by us and not much else. In his version of the future there is no apocalypse, no doom, no gates of hell, no wrath of god or mass hysteria, only sadness. I wonder if perhaps the Age of Loneliness has already begun, its effects far more complicated than we realise.

As I near the end of my trip, I realise I am on the island of last things. There is the last Tasmanian tiger, the last Tasmanian Aborigine, the last Tasmanian emu, the last King Island wombat, the last 'true blue' logger, the last tree, the last Tasmanian devil, the last forest battle. And yet the ghosts of these absences linger. Some are still visible because the land and its people refuse to forget them; they leave empty patches where nothing else can grow. Others have been more myth than reality, adopted as an excuse by authorities to do nothing because it is 'too late,' or taken up as icons for one campaign or another.

If ever there was a canary at the bottom of the world, it is Tasmania, and Wilson's warning about the Age of Loneliness is never far from my mind. I remember leaning on the ferry railing on the way here, trying to will creatures to the surface. I'm not sure I'd like to risk looking at the ocean without that hope.

*

When I return to Camp Florentine, some of the activists have just come back from a fishing trip. Wendy has a box of shucked oysters she chiselled off the rocks. As we sit around the fire, jobs are allocated. People volunteer to sleep in the dragon, or to take over the early-morning watch. Then the gutting of flatheads and the slurping of slimy molluscs gets underway. There's a sense of living like kings out here. The fish are a good feed and slowly people slip off into the forest to sleep.

I'm more relaxed this time. I suspect I've the proprietorial air I accused the other activists of having. There are people here I haven't seen, who don't know the people who used to be here, who don't know me. One day it will irk me that they have no idea whose roles they've stepped into, whose well-worn reins they're holding, whose slogans they've taken to chanting and whose enemies they've adopted without question. But tonight I'm just happy to be camping. There's an empty space behind us. A rubble of road yawning for kilometres, green moss chewed up by yellow machines, and the perfume of broken timber lingering in the air. In the night, dreams are cracked with the sound of falling trees.

The next morning the sound of a saxophone wakes me. I wake up scrambling, thinking it's the emergency horn announcing a police raid. By the time I'm standing on the mud in my thermals, I realise everyone is asleep except me. A tiny tuft of smoke down by the road reveals where one of the boys is on watch. In the forest, a dull-looking grey bird is flitting from tree to tree. It opens its beak and out comes the sound of a saxophone.

I sit down near the fire and start to comb my hair. The coals need something to burn, but I am no good at fires – I always seem to put them out. Some of the blockaders act as if all this is instinctual, as though the knowledge of flint and fire is a long-forgotten reflex. But it's not, not for me anyway. I need to learn how to build a fire. I decide to ask someone to teach me when there is a spare moment. I find myself stabbing at my hair with the comb. Yanking with force, I hit a knot and it painfully breaks off – a tiny dreadlock in the making. I chuck it in the coals and watch it burn.

AFTERWORD TO THE
SECOND EDITION

I kept looking at all the signatures on the page. Perhaps it was naïve of me, but I was entranced by the intimacy, the ballpoint-pen physicality, of each name. Not only had these people – representatives of the timber industry and its workers, and of environmental groups – been in the same room, but they'd *agreed* on something.

For five months, the Wilderness Society, Timber Communities Australia, the National Association of Forestry Industries, the Australian Conservation Foundation, Environment Tasmania, the Forest Industries Association of Tasmania, the CFMEU, the Australian and the Tasmanian Forest Contractors' Associations and the Tasmanian Country Sawmillers' Association sat at the same table and talked. No doubt they cursed one another, laughed at each other's jokes in spite of themselves, perhaps even held doors open for one another – and then they all signed on the dotted line. Could a handwriting analyst looking at this collision of inky squiggles detect the personal, at times deeply bitter, history that connects all of these names?

On 19 October 2010, six weeks after the first edition of this book hit shelves, the forestry roundtable signed off on the Tasmanian Forests Statement of Principles to Lead to an Agreement, a kind of blueprint for future forestry talks. This carefully worded and plainly presented document (no pictures of 'happy working forests' or 'raped landscapes' to be seen) had been no easy outcome. Leaks and rumours dogged the proceedings; at one stage a 'final negotiating draft' was leaked to the press, intimating that environmentalist groups were being pressured to drop their opposition to Gunns' pulp mill in exchange for forest protection. Community groups such as Tasmanians Against the Pulp Mill (TAP), meanwhile, accused the roundtable of being unbalanced and undemocratic – a claim with some validity, considering that representatives of the state's plantation estates were notably absent from the table, as were affected communities.

The leaked draft also revealed one major sticking point in the discussions: the question of 'biomass.' 'Biomass' was essentially a new name for woodchips, only instead of becoming pulp this 'byproduct' would be fed into a massive furnace to generate electricity. The timber industry was touting it as a source of 'renewable energy' – 'renewable' in the sense that 'trees grow back.' Environmentalists, however, weren't having a bar of it. The two sides could either agree to disagree, which would have meant years of new campaigns for and against biomass, or compromise. The roundtable compromised. The Statement of Principles declared that only biomass from plantations could be eligible for Renewable Energy certificates.

The Statement also proposed a moratorium on logging in all high-conservation-value forests (as nominated by the environmental groups) within three months, immediate assistance for timber contractors, an independent assessment of the industry's short- and long-term wood-supply needs, exit and retraining packages for timber workers and an industry-wide transition to plantations, with the exception of a few native-forest sawmills and veneer mills.

All of these recommendations were conditional on federal government funding. Late in 2010, in a revealing union ballot, 97 per cent of Tasmanian members of the CFMEU backed the proposed restructure of the industry. The workers and the environmental groups, it seemed, were finally on the same page.

*

In mid-December 2010, the federal environment minister, Tony Burke, announced that the Gillard government would support the Statement of Principles and appointed Bill Kelty, a former union boss, to broker further negotiations among stakeholders. At a press conference with the Tasmanian premier, Lara Gidding (yes, there is a new Labor premier), Burke said, 'Today's announcement marks the start of two key processes for the next three months: a guaranteed sustainable quantity and quality of wood, and a progressive moratorium on the logging of high-conservation-value forests in keeping with the agreement.'

The moratorium, however, did not happen. Once the fanfare surrounding the announcement subsided, the Tasmanian government began dragging its feet. It took almost four months for the state to instruct Forestry Tasmania to reschedule its logging plans in preparation for a moratorium, and even then it provided the agency with a 'get out of jail free' card, stipulating that FT's first priority was, of course, its contractual obligations. In March 2011, Forestry Tasmania and the state government renegotiated a six-month 'transition period,' during which the agency would phase out its logging in high-conservation-value forests. Activists, however, say they have seen no signs of the agency slowing down.

Eighteen months after the initial Statement of Principles, the blockade at the Florentine forest was still standing. Further south, activists in the Huon Valley watched in disbelief as forests marked for immediate protection continued to fall. Forestry Tasmania had even built new logging roads into previously untouched wilderness. All of which makes it difficult to stave off the onset of cynicism.

Which reminds me: did I tell you that Gunns has stopped logging native forests?

*

In September 2010, Greg L'Estrange fronted a national timber forum and conceded defeat. The greenies, he said, had won.

'The industry has been out-thought and out-played,' he told the Forest Industry Development Conference. 'We have lost the public debate and the support of the broader community ... This may well mean transitioning to plantations, but move we must, for the conflict must end. Too many people have been financially and emotionally injured in the Australian forest wars.'

With more than $600 million of debt, Gunns began selling off assets – even the company's Launceston headquarters was put up for sale. When its Triabunna woodchip mill – the most vital piece of the state's native-forest woodchipping infrastructure – went on the market in June 2011, Aprin, a local father-and-son business owned by Ron and Brendon O'Connor, put in a bid and applied for a multi-million-dollar loan from the Tasmanian government to pay for it. Bizarrely, the government approved the loan. Allegations soon followed that Aprin was acting as a front for Forestry Tasmania and the old native-forest logging guard. Under intense scrutiny, the deputy premier and minister for primary industries, Bryan Green, was forced to admit in parliament that FT would not simply be supplying the woodchip mill if Aprin's bid were successful.

'Yes,' said Green, squirming, 'I can confirm that there is a profit-sharing arrangement … potentially.' What happened next was perhaps the most head-spinning development yet in Tasmania's forest wars.

In mid-June, the O'Connors met journalists in the eastern coastal town of Triabunna and announced that their $16 million dollar bid had been successful. A local ABC radio presenter, Penny Terry, as if wary of jumping the gun, asked the two men if she could really now call them the new owners of the Triabunna mill.

'Yes, yes. Yes, you can,' said Ron O'Connor confidently. 'Signed and sealed.'

The two men went on to espouse the merits of using the 'waste' left on the forest floor, echoing Gunns' old arguments in defence of native-forest woodchipping. This had always been a back of the mind fear for activists. After all, getting Gunns to stop logging native forests would not necessarily stop others from doing so.

A month later, Gunns made a very different and wholly unexpected announcement, sending shockwaves through the timber industry. Gunns had in fact sold the Triabunna mill to two wealthy local environmentalists, Graeme Wood and Jan Cameron, for $10 million – $6 million less than Aprin had offered. Wood, a major donor to the Greens and creator of Wotif.com, was valued at $337 million on the *BRW* rich list in 2011. Cameron, founder of the outdoor clothing company Kathmandu, is estimated to be worth $295 million.

Despite claims by L'Estrange that the decision was purely financial – Aprin hadn't been able to finalise its side of the deal, he said – the O'Connors were bewildered. The Tasmanian government said it was 'disappointed': 'the vegetarians have bought the abattoir,' Premier Giddings observed. Forestry Tasmania accused Gunns of 'blockading' the mill by selling it to environmentalists, effectively preventing competitors from taking advantage of its departure. And for their part, Cameron and Wood announced they had plans to turn the mill into an eco-resort in five years' time, but would re-open the mill in the interim in order to fulfill current timber contracts while the industry restructured.

In selling the Triabunna mill to Wood and Cameron, was Gunns acting as a good corporate citizen or a canny competitor? The company has been busy scrubbing its slate clean, but its reinvention as an environmentally conscious company can go only so far. Its controversial Tamar Valley pulp mill remains in the wings, and is now the one thing standing between the company and non-existence.

When it comes to the mill, Gunns is damned if it does and damned if it doesn't. The company cannot afford to resubmit its proposal to an independent planning body and essentially undo the Lennon government's fast-tracked approval. Yet this is its only hope of obtaining any form of community support.

In an effort to find a way through this dilemma, Gunns has announced tough changes to the mill, some of which feel like déjà vu to Tasmanians, who have already witnessed a merry-go-round of pulp-mill promises. For example, the latest incarnation will use plantation-only woodchips, much like the pulp mill proposed in 2004. Its bleaching technology has been upgraded from 'Elemental Chlorine Free' (ECF) to 'ECF Light.' The 'light' technology is said to reduce chlorine dioxide in the mill's effluent by 40 per cent and create stronger pulp, producing paper that can be recycled several times. However, as Greens leaders have noted, ECF Light is still not in line with Gunns' original promises in 2004, when the company committed to using 'Only world's best technology, utilising a low-impact Total Chlorine Free (TCF) mill.'

This point is key to Gunns' current (self-inflicted) problems. Trust in what the company says has well and truly disintegrated –

and unfortunately for the politicians left to clean up the mess, any policy seen to be helping Gunns is automatically tainted with suspicion. Nevertheless, the mill's perceived economic potential, especially during precarious economic times, has helped to attract both state and federal support.

'Before [the economic crisis] the pulp mill was the icing on the cake, the cream,' Lara Giddings told the *Mercury*. 'Now it is the cake.'

After announcing the tougher environmental protections, Gunns gained an extension and then federal approval for the mill in March 2011. In response, rallies and arrests continue to thwart the project, including an underwater action in which divers held up anti-mill banners amid a protected colony of fur seals in the Bass Strait. Community groups have called for a Royal Commission into the Pulp Mill Assessment Act 2007. In particular, the protesters want to repeal Section 11, the clause they say removes the right of people to claim compensation or take legal action should the mill have a negative impact on their health or livelihood. Court actions against Gunns include allegations that the company has breached its planning permits. The accusation of expired permits has spurred Gunns into action: the mill site is now whirring with action, as earthmoving machines and hi-vis-clad workers dodge activists attaching themselves to machinery.

The company is still without a financial partner for the mill, but L'Estrange maintains it is in 'discussions' with interested stakeholders. Shareholders have been told to expect an announcement soon, while a market update in September 2011 predicted that an investment partner would be announced in 2012, provided the company met an obligation to retire $340 million of primary debt. Meanwhile, as Gunns' workforce is cut back, there are rumours that former employees' loyalty is beginning to sour. More and more dirty laundry has been aired.

In September 2011, thirteen emails were leaked to the *Tasmanian Times*, giving a behind-the-scenes look at the preparation of the pulp mill's impact statement in 2005 and 2006. The communications were mostly between Les Baker, then project manager of the proposed mill, and an employee of Gavin Anderson & Co., an international public relations firm. The PR firm relayed concerns that the 'dark side' (meaning people opposed to the mill) would have an easy time picking holes in Gunns' assessment of the proposed site. In one email, Gavin Anderson & Co. warned that the site-selection documentation was 'significantly deficient': 'We only have some flora and fauna memos, a heritage memo and the very brief desktop report conducted by JP to substantiate the claim that Bell Bay is the preferred site.' In another, Baker ordered the removal of a table showing dioxin concentrations at locations around George Town, saying its inclusion could 'bring down two governments and the company.'

Around the same time, Gunns was spending enormous amounts of money on a media campaign claiming that the mill would be the cleanest and greenest in the world, and that it would meet all of the guidelines set by the Resource Planning and Development Commission. Although Gunns has questioned the authenticity of the emails, Les Baker has confirmed they were his, but adds that they were 'ancient history' and were just some of thousands written in the early stages of the project. Nevertheless, they clearly show that while Gunns was publicly defending its environmental credentials, the company's very own consultants were saying otherwise.

*

Back in the boardroom, the second round of negotiations based on the Statement of Principles was struggling. None of the representatives from Environment Tasmania, the Wilderness Society or the Australian Conservation Foundation would support Gunns' pulp mill, and discussions had ground to a halt. In March 2011, in an interim report submitted to the federal and state governments, Bill Kelty said 'it would be easy to conclude that there is insufficient agreement to establish a workable solution.'

In May 2011, the Wilderness Society pulled out, calling on the government to implement a logging moratorium and financial assistance for timber contractors. Neither measure was introduced and the talks, minus one, limped on.

On 22 June, the roundtable delivered an agreement. A month later, Prime Minister Gillard and Premier Giddings announced a 'Heads of Agreement' to implement the roundtable's recommendations. Two weeks later, Gillard and Giddings signed the Tasmanian Forests Intergovernmental Agreement.

Under the agreement the federal government announced a $276 million funding package, including $15 million from the Tasmanian state government. Of this money, $85 million would be used to help contractors affected by the downturn in the industry (in particular those affected by Gunns' decision to cease native-forest harvesting); $43 million would facilitate the protection of new areas of high-conservation-value forests; $120 million over fifteen years, including an initial payment of $20 million, would go towards identifying and funding regional development projects; and $7 million per year on an ongoing basis would fund the management of new reserves. The agreement also declared that 430,000 hectares of forest in areas nominated by environmental groups – including the Florentine Valley, the Weld, the Tarkine, Wielangta and Blue

Tier – were to be set aside for 'immediate' interim protection while an independent analysis of their conservation value was conducted.

The next sticking point was: how much compensation should Gunns receive for giving up its twenty-year pulpwood agreement with Forestry Tasmania? The two major wood-supply contracts were valued at $200 million per year when they were signed in 2007 and it was widely reported that Gunns was seeking a $100 million payout. The impoverished Tasmanian government had no means of paying such a sum and besides, Forestry Tasmania was claiming that Gunns owed it $25 million. Eventually, after a series of paltry offers, the state opted to divert funds from the forestry agreement package to settle the dispute. Using money originally set aside to manage new forest reserves and consult regional communities, $23 million would go to Gunns and $11.5 million to Forestry Tasmania.

Understandably, many observers – especially timber workers, who were yet to receive any assistance – were not impressed. Most green groups, however, felt that compensating Gunns and retiring Gunns' wood entitlements so they could not be on-sold was a necessary step to safeguard the peace deal. They reiterated that the compensation was to pay for the entitlements of 107 workers made redundant from Gunns' five native-forest sawmills. Giving money to Forestry Tasmania, however, seemed plainly absurd to the Wilderness Society, especially when FT was seeking ongoing access to the newly protected areas.

'Forestry Tasmania continues new logging and roading in forests that should be protected. Providing them with an $11.5 million blank cheque will do nothing for forests needing immediate protection,' the Wilderness Society's Vica Bayley said.

Forestry Tasmania has insisted that it needs to log forty-one coupes inside the protected forests in order to fulfill its existing contracts (predominantly with the Malaysian logging corporation Ta Ann). FT's managing director, Bob Gordon, says the agency is acting within the terms of the Intergovernmental Agreement, which acknowledges that FT has the right to meet its pre-existing commitments. But others argue that FT has had more than enough warning to relocate this logging to less ecologically significant forests. Prior to the signing of the Heads of Agreement, Professor Jonathon West, the independent expert leading the verification process, made a submission to government stating that after reviewing FT's publicly available three-year harvesting plan for the financial year to June 2012, his view was that they could source adequate high-quality sawlogs from outside the interim protected area.

However in a letter to the state government, FT claimed it could not reschedule 'due to planning and operational constraints such as restricted access during the wedge-tail eagle breeding season and lack of suitable roaded coupes.'

Independent 're-schedulers' have now been appointed by the government to rejig Forestry Tasmania's logging schedule. In the meantime, the state agency is continuing to plough through numerous coupes earmarked for interim protection, including parts of the Tarkine rainforest. It is amazing how quickly an assessment can sacrifice a forest and how slowly an assessment can protect one.

During the very first roundtable discussions, Ross Story from Tasmanians Against the Pulp Mill told the ABC he was concerned that Forestry Tasmania was not at the initial talks and therefore would not be part of any formal agreement.

'In future,' he said, 'they may well pursue their own agenda and that could include operating woodchip plants themselves as well as wood-fired power stations.'

That is exactly what Forestry Tasmania has attempted to do. From backing Aprin's bid for the woodchip mill to refusing to relocate its logging, the state agency has shown blatant disrespect for the agreement. Furthermore, environment groups claim that FT is blocking access to environmental and mapping data, making the reserve agenda difficult to implement. That the independent committee assigned to assess the potential reserves will have to work closely with FT and its data has the green groups nervous. For many local forest activists, the enemy of the forests, once Gunns, is now Forestry Tasmania.

Given this, the hoped-for peace in Tasmania's forests remains elusive. The movers and shakers on both sides are increasingly tight-lipped and white-knuckled, bent over a high-stakes game of Chinese checkers in which Tasmania and its forest industry will receive $276 million, and yet there is no guarantee that any forest areas nominated to be assessed will be put into permanent reserve.

*

And so, when it comes to the situation in Tasmania's forests, nothing and everything has changed. John Gay, the former Gunns' boss, is back in the game despite being hauled up to the Magistrates Court on charges of insider trading. It would have been naïve to think he was going to fade into the twilight of retirement after a shareholder revolt ousted him. In October 2011 Gay headed up a small consortium of buyers and purchased a sawmill, a veneer mill and a timber drying & finishing plant from Gunns. Gay's recent activities, however, may well be hampered by his upcoming court dates.

The Australian Securities and Investments Commission is alleging that Gay was privy to inside information when he disposed of 3.4 million shares for between 88 and 92 cents a piece in December

2009. In February, Gunns publicly announced a 22 per cent profit slump, causing its shares to dive to 55 cents each. If he is convicted, Gay will become the most senior Australian executive to be found guilty of insider trading.

As for the 'vegetarians', Jan Cameron and Graeme Wood, on purchasing the Triabunna woodchip mill they stated that they were willing to do whatever was required to ensure the genuine success of the forests peace deal. This was echoed by the newly appointed general manager of their mill, Alec Marr, who also issued a warning to all loggers that once the mill was re-opened, every pulp log load would be scrutinised at the gate. For readers managing to keep up and connect the dots, yes, this is Alec Marr the former director of the Wilderness Society and number one defendant in the Gunns 20 case.

So far, the woodchip mill has not re-opened to logging trucks. Instead a sole employee wanders around the mill and its still machinery, flicking various switches and twisting odd dials, maintaining signs of mechanical life.

November 2011

ACKNOWLEDGMENTS

This book would not even have occurred to me without Ula Majewski. Through meeting Ula and my subsequent friendships with Warrick Jordan, Gemma Tillack and Jess Wright, these four fiercely intelligent individuals provided me with endless chewing tobacco for the mind. I've loved every minute of agreeing *and* disagreeing with you all. Thank you for making each journey of mine to Tasmania feel a kind of homecoming.

My thanks to Lindsay Tuffin, a brilliant praying-mantis of a man, creator and editor of the *Tasmanian Times*, a news website described by former premier Paul Lennon as 'fucking useless.' I beg to differ. To Dave Obendorf, whose research, eloquence and enthusiasm have been a special touchstone for me – I hope your tadpoles have grown up and no longer occupy your bathtub. So many Tasmanians gave me their time that I cannot list them all here – but I do need especially to thank Frank Nicklason, Geoff King, Bruce Montgomery, Prue Barratt, John Alomes, Jack Lomax, Bill Manning, Michael Field, Pete Hay, Matthew Newton, Kevin Perkins, Miranda Gibson and Bridget Gatenby.

Then there are the Tasmanians who took me in. The Johannsohn family, who never blinked when they returned home to find me in their lounge room, curled up with the cat, or when I started to refer to their spare room as my own. Again, there are too many people to list individually, but many thanks to those who welcomed me, fed me, made me sandwiches for the road, drove out to meet me when a bridge flooded, and so on. Thank you all for showing me there was, and is, more to Tasmania than trees.

Also, the talented Laura Minnebo, whose photograph is on the cover of this book, and Amanda Lohrey, the fine and formidable writer whose *Quarterly Essay Groundswell* compactly and succinctly captured the rise of the Australian Greens – thank you, Amanda, for your feedback in the course of writing this book.

On the mainland: A big thank you to my editors, Chris Feik and Denise O'Dea, who transformed my manuscript – which I fear may have initially resembled Gunns' pulp-mill submission, heavily criticised for reading like a 'dog's breakfast.' To Alan Attwood and his editorial team at the *Big Issue*, who took on my original 20,000-word essay on Tasmania's forests, laboured tirelessly on the unruly beast and dedicated a whole issue to the story.

To my 'one phone-call,' Benjamin Law, whose general collapse into hysterics at my Tasmanian failures seemed to make everything okay and surmountable (gastro included). Thanks to my mum for the vitamins, and finally, Emilio, whom I told I wouldn't be able to thank because there wouldn't be enough room. He replied, 'Well, make some,' and he was right. Thanks, Em, and sorry about the mess.

www.ingramcontent.com/pod-product-compliance
Lightning Source LLC
Chambersburg PA
CBHW052121270326
41930CB00012B/2708